THE NEUROSCIENCE OF EMOTION

The Neuroscience of Emotion

A NEW SYNTHESIS

Ralph Adolphs and David J. Anderson

Princeton University Press   Princeton and Oxford

Published by Princeton University Press
41 William Street, Princeton, New Jersey 08540

In the United Kingdom: Princeton University Press
6 Oxford Street, Woodstock, Oxfordshire OX20 1TR

press.princeton.edu

Jacket images courtesy of The Noun Project (thenounproject.com).
Human head icon (modified) by Gregor Cresnar.
Dog icon (modified) by Boris Farias.

ISBN 978-0-691-17408-2

Library of Congress Control Number: 2018933553

British Library Cataloging-in-Publication Data is available

This book has been composed in Adobe Text Pro and Replica Pro

Printed on acid-free paper. ∞

Printed in the United States of America

10 9 8 7 6 5 4 3 2

CONTENTS

List of Illustrations   ix
Preface   xi
Acknowledgments   xv

PART I. Foundations

CHAPTER 1. What Don't We Know about Emotions?   3
Emotions According to *Inside Out*   4
Toward a Science of Emotion   13
Emotions Are Decoupled Reflexes   18
Questions We Will Not Answer in This Book   23
What Do We Want to Know about Emotions?   25

CHAPTER 2. A Framework for Studying Emotions   29
Warm-Up: Neuroscience Questions about Emotion   30
Toward a Functional Definition of Emotion   39
Proper Functions and Malfunctions   43
Emotions and Consciousness   49
An Experimental Example   52
Summary   57

CHAPTER 3. Building Blocks and Features of Emotions   58
Building Blocks versus Features   62
A Provisional List of Emotion Properties   65
Summary   98

PART II. Neuroscience

CHAPTER 4. The Logic of Neuroscientific Explanations   103
Levels of Biological Organization   105
The Concept of Mechanism in Neuroscience   108
Testing Causal Relationships between Neural Activity and
     Behavior   114
Levels of Abstraction   115

Mixing of Terms in Neuroscience Explanations   120
Necessity, Sufficiency, and Normalcy   123
Summary   125

**CHAPTER 5. The Neurobiology of Emotion in Animals:
General Considerations**   127
Why Do We Need Studies of the Neurobiology of Emotion in
   Animals?   131
What Do We Want to Understand about Emotion by Studying
   Animals?   140
The Relationship of Emotion States to Motivation, Arousal,
   and Drive   143
Psychiatric Drugs, Animal Models, and Emotions   153
Summary   159

CHAPTER 6. The Neuroscience of Emotion in Rodents   161
Emotion, Fear, and the Amygdala   163
Innate Defensive Behaviors and Emotions   179
Distributed versus Localized Emotions in the Brain   188
Anxiety   189
Other Emotion States: Aggression and Anger   193
Positively Valenced Emotion States   195
Summary   196

CHAPTER 7. Emotions in Insects and Other Invertebrates   197
Learned Avoidance Behavior in *Drosophila*   198
Does *Drosophila* Have Emotion States?   203
Anxiety in Insects and Other Arthropods   208
Emotion States and Social Behavior in Insects   209
Internal States in Other Invertebrates   210
Summary   214

CHAPTER 8. Tools and Methods in Human Neuroscience   215
Historical Neuroscience Studies of Emotion in Humans   218
fMRI Studies of Emotion: The Method   233
Similarity Analyses   242
fMRI in Animals?   245
Summary   249

CHAPTER 9. The Neuroscience of Emotion in Humans 251

fMRI Studies of Emotion: The Logic and the Challenge 251

Lessons from Two Examples: Music and Faces 253

Attributing Emotions to Others 257

Imaging Emotion Concepts 260

Feeling Emotions 265

Central Emotion States 270

Dissociating Emotion States from Concepts and Experience 273

Summary 278

PART III. Open Questions

CHAPTER 10. Theories of Emotions and Feelings 281

The Structure of Affect 282

Theories of Feelings 286

Philosophy of Emotion 299

Taking Stock 303

Summary 306

CHAPTER 11. Summary and Future Directions 308

The Main Points of This Book 308

Feelings Again 311

Future Experiments 313

Glossary of Terms and Abbreviations 327

References 337

Index 347

ILLUSTRATIONS

1.1.	MRI scan of a mother kissing her child	14
1.2.	Emotions can be inferred from several kinds of data	17
1.3.	Functional varieties of disgust	21
1.4.	Desiderata for a science of emotion	26
2.1.	Orchestration of many components of a fear response by the amygdala	34
2.2.	Bodily maps of people's beliefs about emotional feelings	36
2.3.	Emotions are functional states	41
2.4.	A detailed functional architecture for threat processing	54
2.5.	Response of a mouse to a looming stimulus	56
3.1.	Spatiotemporal extent of an emotion state	59
3.2.	The properties of emotion states	66
3.3.	A dimensional space for emotions	68
3.4.	Drift-diffusion model	74
3.5.	Causal models	82
3.6.	Dual-process theories of cognition	84
3.7.	Emotion regulation as a metacognitive ability	88
3.8.	Expression of emotions in animals	91
4.1.	Different scales of biological organization in the brain	107
4.2.	Methods for investigating the brain at different scales	109
4.3.	Correlation versus causation	112
4.4.	Levels of abstraction for investigating emotions	117
5.1.	Different ways of thinking about emotions and MAD states	145
5.2.	Konrad Lorenz's hydraulic conception of drive	149
5.3.	Two alternative explanations for the action of anxiolytic drugs	155
6.1.	Pavlovian fear conditioning	166
6.2.	Fear conditioning and amygdala nuclei	170
6.3.	Neuronal subpopulations in the central amygdala	173
6.4.	Resolution and cellular specificity of methods to measure amygdala activity in humans versus mice	175
6.5.	Distinct brain regions involved in different types of fear	184

6.6. Behavioral assays and brain networks for fear and anxiety in rodents 192

7.1. Compartmental release of dopamine controls appetitive versus aversive conditioning 202

7.2. Experimental system for investigating whether innate defensive responses in flies display any emotion primitives 204

8.1. Brain regions commonly studied in human emotion 224

8.2. Amygdala lesions impair the processing of information from the eye region of faces 228

8.3. Meta-analytic mapping of brain activations for emotion 232

8.4. Representational similarity analysis 244

8.5. fMRI in dogs 249

9.1. Attributing emotions and their causes 258

9.2. Brain regions involved in emotion attribution 259

9.3. Representational similarity analysis of emotion concepts 261

9.4. Inducing strong emotion states in an fMRI study with realistic stimuli 271

9.5. Threat imminence activates specific brain regions 272

9.6. Similarity between the structure of different emotion aspects 275

10.1. Feelings in everyday life 285

10.2. The theory of constructed emotion 292

11.1. Hypothesis for how the amygdala produces components of fear 318

11.2. Combining optogenetics with fMRI 323

11.3. Richard Feynman's blackboard at the time of his death 326

PREFACE

Emotions are one of the most apparent and important aspects of our lives, yet have remained one of the most enigmatic to explain scientifically. On the one hand, nothing seems more obvious than that we and many other animals have emotions: we talk about emotions all the time, and they feature prominently in our literature, films, and other arts. On the other hand, the scientific study of emotions is a piecemeal and confused discipline, with some views advocating that we get rid of the word *emotion* altogether. If you ask scientists, even those in the field, what they mean by an emotion, you will either get no explanation at all or else several quite discrepant ones that seem to be referring to quite different phenomena. We aim to provide a fresh look at emotion from the perspective of biology, a perspective that can provide a foundation for the field from which to move forward in a productive, cross-disciplinary fashion.

Emotions and feelings have been the topic of countless books, some of them detailed technical books (often a collection of chapters from many different authors), and most of them popular books focused on the psychology of emotion. Ours is none of these. It is not intended as a textbook, a popular book, or a monograph of any sort. Instead, our aim in writing this book was to take stock of the field, from a fairly high-level perspective, to provide a survey of the neurobiology of emotion, and, most importantly, to provide both a conceptual framework and ideas for approaches that could be used by a neuroscience of emotion going forward. Our intended audience is any educated reader, but our core audience is students and researchers who are contemplating going into the field of affective neuroscience, or who are already in the field and wondering what path their research should take. We also hope that at least a good part of the book would be accessible and interesting to readers who do not have a strong scientific background. Indeed, it is entirely possible to glean most of the conceptual framework just from reading chapters 1–4 and chapter 11, and skipping some of the more detailed chapters in the middle. We decided to eschew detailed citations of the papers behind every point and study that we describe, instead choosing to give a more broadly accessible treatment that only cites the most important key papers or reviews (which, in turn, will provide interested readers with a longer list of further references).

Our book differs from most other books on emotion in scope and organization. One of us (Adolphs) investigates emotion in humans; the other (Anderson) investigates emotion in mice and flies. This breadth of different backgrounds, and the presentation of the different species studied, is a critical ingredient of this book, since it forced us to abstract from many details in order to uncover fundamental principles that would cut across different approaches and different species. It also meant that neither one of us is in fact the authority for all of the book: notwithstanding extensive discussions, comments, and cowriting, there are parts of the book that have only one of us as the principal author and expert. Indeed, there are parts of the book on which we continue to disagree!

We do not intend to provide a comprehensive new theory of emotion. Indeed, we don't feel that we provide any kind of theory of emotion at all. Instead, we describe ways that scientists should think about emotion, and ways that they should use the word *emotion* consistently in their science, in order to forge a neuroscience of emotion with the maximal long-lasting impact. Our intent was to provide a framework for investigating emotions that would be applicable to those working in animal models; those working with human subjects; those using functional magnetic resonance imaging (fMRI), electrophysiology, optogenetics, or clinical populations. We even hope that what we have written here would be useful to engineers who are trying to figure out how to build robots that have emotions. In our view, a science of emotion needs to meet two criteria: it should be comprehensive and it should be cumulative.

Forging a comprehensive science means that the encapsulation often evident in papers, journals, and meetings on emotion needs to be overcome. Scientists studying emotion in rats and in humans need to be able to speak to one another, rather than build walls that isolate their research enterprise from the rest. A comprehensive science of emotion also needs to connect with all domains of science that are relevant to emotion: it needs to connect with psychology and with neurobiology. Doing this requires a consistent terminology that makes principled distinctions, and that allows clear operationalization of the different concepts that a science of emotion will use. We spend some time in the first three chapters articulating such distinctions and outlining the features of emotion that a scientist would look for, whether she is studying emotion in humans, rodents, or flies. This approach necessitates some terminological commitments, and we explain these in the early chapters. We also return to them when we compare our view to some of the many other theories of emotion out there, in chapter 10.

A high-quality science of emotion requires not only clear terminology and operationalization of concepts, it also requires sensitive measures, statistically robust analysis tools, and creative hypotheses. The later chapters take up these issues in the context of a survey of ongoing neuroscience studies. Taken together, these ingredients would enable a cumulative science of emotion, a science in which current studies can build on prior work, and in which the accumulation of many studies over time allows comparisons and contrasts, as well as syntheses and formal meta-analyses. We are currently a long way from having achieved this. Indeed, most meta-analyses of emotion are either extremely narrow, or else hopelessly inconclusive because they mix studies with very different standards or terms.

There is no question in our minds that emotions are real phenomena that need to be explained. We believe that, in addition to humans, many other animals have emotions—both of the authors of this book have cats as pets, and we are convinced that they have emotions. However, intuition and belief are not the same as scientific knowledge, and an important goal of the book is to suggest objective criteria to apply in searching for cases of emotional expression in animals. Finally, we also believe that emotions are states of the brain, and that the mechanisms that generate emotions can be investigated with neurobiology. Our book is based on these underlying assumptions; we summarize them again in the very last chapter.

ACKNOWLEDGMENTS

We thank Daniel Andler, Lisa Feldman Barrett, Andrea Choe, Antonio Damasio, Karen DiConcetto, Frederick Eberhardt, Yanting Han, Prabhat Kunwar, Joseph LeDoux, Uri Maoz, Aaron Mendelowitz, Leonard Mlodinow, Amanda Nili, Oana Tudusciuc and Moriel Zelikowsky for helpful comments and discussion on drafts of the book.

Special thanks are due to Len Mlodinow and Amanda Nili, who provided detailed edits on parts of the book; to Anna Skomorovsky, who made most of the figures; to Yuki Toy, who created sketches of humans and animals; and to Sheryl Cobb, who obtained copyright permissions. We are grateful to Dawn Hall for expert copyediting, and to Alison Kalett and the staff of Princeton University Press for their comprehensive support in all aspects of production and marketing.

Ralph Adolphs would like to thank Tetsuro Matsuzawa, Satoshi Hirata, and the Primatology and Wildlife Science graduate program at Kyoto University for their generous hosting of a brief sabbatical in Kyoto in 2016, during which this book was conceived. I also thank my graduate and postdoctoral mentors for the role they played in shaping much of the approach to studying emotions presented in this book. My PhD advisor, Mark Konishi, was steeped in neuroethology and always stressed the importance of relating any experimental finding to aspects of an animal's natural environment in which a behavior or brain function served some adaptive role. My postdoctoral mentor, Antonio Damasio, was working on theories of emotions and feelings when I first began my postdoctoral work, and these had a large influence on how I think about emotion, and shaped the direction of my subsequent career. Finally, I would like to thank my family both large and small; my lab for their questions, criticisms, and sustained contributions to affective neuroscience over the years; and my wife and parents, without whose support and understanding I would not be doing science at all.

David J. Anderson would like to acknowledge the influence of his colleague, the late Seymour Benzer, for encouraging him to begin to investigate emotion states in flies, and Martin Heisenberg and the Alexander von Humboldt Society for supporting a sabbatical in his laboratory at the University of Würzburg, Germany, during which many of the ideas presented in this book were discussed and preliminary experiments with flies on fear were conducted. I would also like to thank my postdoctoral mentor, Richard Axel, for encouraging my switch to the field of emotion from my original study of stem cells and neural development. Finally, I would like to thank members of my laboratory past and present for their contributions to the research discussed, and for critical feedback on the ideas presented here. Most importantly I would like to thank my wife, Debra, for her love and patience with my own "mad pursuit" (Crick 1988).

We both thank all our students and colleagues at Caltech for innumerable discussions from which this book benefited. Caltech's intense and cross-disciplinary intellectual environment played a major supporting role in generating much of what is distinctive about this book: an emphasis on questions over answers, and on an integrated and quantitative scientific approach.

We dedicate this book to the memory of Jaak Panksepp (1943–2017), who made important contributions to the neuroscience of emotion, and whose own work in animals highlighted the importance of subcortical structures in mediating specific emotions. His book, *Affective Neuroscience: The Foundations of Human and Animal Emotions*, remains a landmark in the field.

PART I

Foundations

CHAPTER 1

What Don't We Know about Emotions?

The greatest enemy of knowledge is not ignorance,
it is the illusion of knowledge.
—Ralph J. Boorstin

If you are like most people, you feel convinced that, because you have emotions, you know a lot about what emotions are, and how they work. We believe you are almost certainly wrong. In the field of emotion, as in most fields, familiarity is not the same as expertise. After all, you have a heart, but that doesn't make you an expert on hearts. You leave that to your cardiologist.

Yet the science of emotion is fraught with this problem: everyone seems to think they know what an emotion is. This is because we all have strong, and typically unjustified, intuitive beliefs about emotions. For instance, some people are absolutely certain that animals have emotions; others are absolutely certain that animals could not have emotions. Neither camp can usually give you convincing reasons for their beliefs, but they stick to them nonetheless.

We cannot emphasize enough the pervasive grip that our common-sense view of emotions has on how we (that is, researchers in the field) frame our scientific questions. We need to free ourselves of our commonsense assumptions—or at least question all of them—if we want to ask the right questions in the first place. This chapter introduces the topics of this book through this important premise and concludes by listing what we ideally would want from a mature science of emotion, and what entries in this list we will tackle in this book.

We wrote this book for two overarching aims. The first aim is to motivate the topic of emotion, to note that it is of great interest not only to laypeople but also to many scientific fields of study, and that it is a very important topic as well. At the same time, we emphasize that we currently know remarkably little about it yet—in particular, we know a lot less than we think we know. This is good news for scientists: there is work to be done, interesting and important work.

The second aim is to provide a summary of what we do know and to sketch a framework within which to understand those empirical findings and within which to formulate new questions for the future. This process is in practice very piecemeal: we need to have a little bit of data even to begin thinking about what emotions are, but then we discover problems with the way prior experiments were done and interpreted. In the dialectic of actual scientific investigation, both conceptual framework and empirical discovery are continuously revised, and inform each other. However, we have not written our book this way. Instead, we begin with some of the foundations for a science of emotion (chapter 2)—what kinds of ontological and epistemological commitments it requires, what kind of structure an explanation takes—and then work our way toward a list of features or properties of emotions (chapter 3), which then finally are the things we look for, and discover, through empirical research (chapters 4–9). We return to the foundations and the questions again in chapters 10 and 11 by contrasting our views with those of others, and by suggesting some experiments for the future.

Emotions According to *Inside Out*

What is it about emotions that we would like to understand? And what do we *think* we understand, but in fact don't (or are mistaken about)? Because emotions are ubiquitous in our lives, and integral to our experience of the world, it is dangerously easy to come up with simplistic views that do not stand up to closer scrutiny, and instead impede scientific progress because they create "the illusion of knowledge."

The film *Inside Out*, which won the 2016 Academy Award for Best Animated Feature, as well as a Golden Globe, provides a good example of many common but incorrect assumptions about emotion. As you watch the film, you get a fanciful view of how emotions are supposed to work inside a twelve-year-old girl, how those emotions are supposed to be integrated with memory and personality, and how they are supposed to be expressed as behavior. If *Inside Out*'s view of emotion were right, you would be tempted to conclude that we understand an enormous amount about how emotions work—and, more generally, about how the mind and brain work. But *Inside Out*'s view of how emotions work is wrong. In examining what, exactly, *is* wrong with it, we can highlight some of the gaps in our current understanding of emotion. If you've seen the film and you already find the view of emotion portrayed by *Inside Out* silly, you are ahead of the game—but bear with us as we use it as an example for uncovering problematic beliefs about emotion.

Inside Out's view of emotion takes as its starting premise the idea that all our emotions boil down to a few primary ones: in the film, they are joy, anger, fear, sadness, and disgust. These five emotions are animated as different characters, charming little homunculi that live in the brain of the little girl and fight with each other for control of her behavior and mental state. These homunculi sit at a control panel and watch the outside world on a screen. They react to the outside world, and in response they manipulate levers and switches that control the little girl's behavior. They are also affected by memories that are symbolized by transparent marbles; moreover, a series of theme parks provide a mental landscape symbolizing different aspects of the girl's personality. The five emotion characters fight over access to the memory marbles and struggle to keep the girl's theme-park attractions open for business.

From the film's point of view, the five emotions are the dominant force controlling the little girl's thoughts, memories, personality, and behavior; thinking, reasoning, and other cognitive activities are relegated to a sideshow. Truly, the little girl is an entirely emotional being. These details of the movie may not represent the way you think about emotions, but they characterize how many people do.

So what's wrong with the film's creative, engaging metaphor? Let's unpack a few of the key ideas about emotions that *Inside Out* showcases, highlight the errors in their underlying assumptions, and try to articulate the scientific questions that they raise. Although science may not yet have the answers, the exercise will help us frame the issues.

Idea 1. There are a few primary emotions. The prevailing view, enshrined in many psychology textbooks, is that there is a small set of "primary" or "basic" emotions: as we already mentioned, these are joy, anger, fear, sadness, and disgust, according to *Inside Out*. Different scientific emotion theories offer a big range in the number of basic emotions—anywhere from two to eleven! A second type of emotion is often called "social" or "moral" emotion and typically includes shame, embarrassment, pride, and others. These social emotions are thought to be more essentially tied to social communication than the basic emotions are. But although there are multiple schemes, many classic emotion theories tend to share the idea of a fixed, and relatively small, set of emotions that correspond to the words we have for emotions in English.

The idea of a small set of basic emotions was most notably introduced by the psychologist Paul Ekman, based largely on data from his studies of emotional facial expressions in humans. Ekman argued that facial expressions of basic emotions can be recognized across all human cultures (Ekman 1994); he studied them even among tribes in New Guinea. Ekman's set of basic emotions includes happiness, surprise, fear, anger, disgust, and sadness (although contempt is also sometimes included). The neurobiologist Jaak Panksepp similarly proposed a set of basic emotions, derived from his observations of animal behavior: seeking, rage, fear, lust, care, panic, and play (Panksepp 1998). These emotion theories have much to recommend them and stimulated entire lines of important research. But they also suggest two questionable background assumptions (which Ekman and Panksepp themselves may or may not have held).

Questionable assumption 1: Emotions (at least the "primary" ones) are irreducible. A presumption that often accompanies the idea of a small set of primary emotions is that they are irreducible units. According to this assumption, emotions like "fear" or "anger" cannot be broken down into further components that are still emotional. The psychologist Lisa Feldman Barrett has argued strongly against this assumption, pointing out that it requires belief in some kind of mysterious "essences" of emotions—the belief that there is something irreducible that makes each primary emotion the emotion that it is (Feldman Barrett 2017a). This central assumption underlies the representation of each of the primary emotions in *Inside Out* as a distinct *character*.

"Joy" and "fear" do not merge with each other; they are each unique individuals. They have stable, fixed identities and functions, and do not share components (for example, in the movie's metaphorical language, they do not share internal organs, limbs, and such).

Yet there is scant scientific evidence that "joy," "fear," or "anger" are irreducible and do not share component parts. Equally plausible is an alternative view in which each of these emotions is made up of a collection of components, or building blocks, some of which are shared by other emotions. Initial doubts such as these lead to the following set of scientific questions that can serve as a starting point for further investigation:

"Are different emotion states composed of features or dimensions that are shared, to variable extents, across multiple emotions? Are some emotions composed of, or based on, combinations of other more basic emotions?"

Questionable assumption 2: the primary emotions correspond to those for which we have names in English.

Related to questionable assumption 1 is the idea that words like "happiness," "fear," "anger," and so forth in fact pick out scientifically principled categories of emotion. It is easy to see why this is unlikely to be the case. For one, we had these words for emotions long before there was any science of emotion—so why would one expect them to align well with scientific emotion categories? For another, different cultures have different words for emotions, and many of these turn out to be extraordinarily difficult to translate. In German, the word "Schadenfreude" denotes the emotion we feel when we feel happy about somebody else's misfortune. Should that be a primary emotion, just because there's a common word for it in German? There are many more such examples, entertainingly cataloged in Tiffany Watt Smith's book, *The Book of Human Emotions* (Smith 2016). This poses some important scientific questions:

"How should we taxonomize emotions? How many emotions are there, and what names should we give to them? Are there different emotions in different cultures? Are there different emotions in different species? Can we use a word like 'fear' to refer to the same type of emotion state in a person, a dog, and a cat? How and when in evolution did emotions first arise, and how did they diversify?"

Given how little we yet know about these questions, and given that there are good reasons to believe our current emotion categories ("happiness," "sadness," and such) will need to be revised, we will say little in this book about specific emotions. We will refer to some emotions (notably "fear") by way of example. And we will sketch how a future science of emotion might give us better categories or dimensions by which to taxonomize emotions. But this book is primarily about emotions in general, not about specific emotion categories.

Idea 2. Emotions are rigidly triggered by specific external stimuli. In the film *Inside Out*, all five emotion characters sit lazily around the control panel watching a screen that projects the outside world into the little girl's mind, and are aroused into action only when an appropriate stimulus or circumstance appears. In the film, some stimuli do not activate a given emotion at all (for example, the "anger" character often sits dozing in his chair and does not react unless something maddening happens to the girl), while other stimuli activate multiple emotions. If the depiction from the film were accurate, we could easily figure out the emotion states of other people (and presumably other animals) by a straightforward list of rules that link specific stimuli to specific emotions in a characteristic and inflexible manner. This picture assumes that emotions are far simpler and more automatic than we in fact now know them to be. According to Idea 2, emotions would be just like reflexes. Some things will make you happy, others will make you sad, and some will trigger a specific mix of emotions, according to a set of rules.

Questionable assumption 3. Emotions are like reflexes. The movie gets it right that emotions are often triggered by stimuli in our surroundings. But what determines *which* emotions are triggered by *which* stimuli and under *which* circumstances? Why would seeing a dog trigger only a minimal emotional response in some people, and strong fear or happiness (emotional responses of opposite valence) in others? What accounts for the extraordinary flexibility with which many different stimuli, depending on the context and depending on the person, can elicit emotions? One can pose the following scientific questions:

"What determines whether an external stimulus will evoke an emotion or not, and what determines the kind of emotion evoked? What role do development and learning play in determining an organism's response to a given stimulus? How does this process differ from simpler stimulus-response mappings, such as a reflex?"

Idea 3. Emotions control our behavior. The film portrays the emotion characters as controlling the little girl's behavior by operating joysticks on the control panel. The little girl is but a hapless puppet, with emotions determining her behavior. This central visual metaphor encapsulates the title of the movie: our behavior is controlled, from the "inside out," by our emotions. This feature is the counterpart to 2 above, with respect to the behavioral output rather than the stimulus input.

Questionable assumption 4. Specific emotions cause fixed and specific behaviors. Our subjective experience of emotion leads to the intuition that our emotions cause our behavior: I cry *because* I feel sad. Yet not all emotion theorists agree with this assumption. Indeed, the nineteenth-century American psychologist William James argued, counterintuitively, that emotions are a *consequence*, not a cause, of behavior: I feel afraid *because* I run from the bear, I do not run because I feel afraid (James 1884). Yet James already had doubts that just observing bodily reactions was sufficient to identify specific emotion categories. If it were true that specific emotions cause fixed and specific behaviors, we could infallibly deduce a person's emotions just from watching their behavior. If so, then taken together with questionable assumption 3, we wouldn't need emotions at all to explain behavior, there would simply be a set of rules linking stimuli to behavior. That was the view that behaviorism advocated in the earlier twentieth century. One reason for the demise of behaviorism was that people realized that mappings from stimuli to behavior were far too complicated, and too dependent on context, inference, and learning, to be formulated as rules. Emotions, in our view, are internal states that afford a flexible mapping to behavior, as we will detail throughout this book. This leads to the following scientific questions:

"Do internal emotion states cause behavior, or are they merely an accompaniment to behavior? Or might emotions actually be a consequence of behavior? What exactly are the causal links between stimuli, emotions, and behavior? How could we identify emotions in the absence of behavior? After all, we can be angry without punching somebody or showing any other easily detectable behavior."

Idea 4. Different emotions are located in different, discrete brain regions. The beguiling picture of emotions as walking, talking cartoon characters in *Inside Out* is closely aligned with the belief that different emotions must correspond to anatomically distinct modules in the brain. Is there a place in the brain for fear, for example? This is a question that has received a lot of attention, including serious scientific investigation!

Questionable assumption 5. Specific emotions occur in specific brain structures. The era of functional neuroimaging with f MRI, as well as the study of patients with focal brain lesions, has led to the idea that emotions are generated in localized brain structures. For example, findings on the amygdala (a brain region studied in both of our laboratories to which we will return in some detail in later chapters) have led to the popular view that "fear is in the amygdala." Yet more recent work clearly shows that this view cannot be right; indeed, that it does not even make sense, and that emotions depend on a much more distributed set of brain regions. This leads to the following scientific questions:

"How is the processing of emotion carried out across the brain? Are there identifiable functional neural substrates that organize or implement specific emotion states? Or is any given emotion state produced in such a highly distributed manner that it is impossible to assign a function in emotion to any brain region or neuronal cell population? Would it ever be possible to predict what emotion an individual is experiencing purely by examining activity in his/her brain?"

As we will explain later, modern neuroscience approaches have given us a view of brain function that reconciles a dichotomy inherent in these questions. It will turn out that there are no macroscopic brain structures dedicated specifically to emotions (fear is not "in the amygdala"), but that there *is* specificity nonetheless. The specificity is at the level of circuits and cell populations, a level of organization that requires modern neuroscience tools to visualize. We spend some time in chapters 4 and 5 explaining these neuroscience tools, since their logic is required to reformulate the questions about emotion.

Idea 5. Emotions are conscious homunculi. The movie illustrates beautifully the idea that the brain is a machine with a little person (or persons) inside, who views the outside world, reacts to it, and then transfers those reactions to us. In other words, our subjective experience of emotion is created and embodied by the subjective experience of a miniature version of ourselves in our brain, a so-called homunculus (box 1.1). (As an aside, it is also interesting that this view, of little emotion homunculi within ourselves, to some extent relieves us of full responsibility for our emotional behavior—as when we say, "my anger made me do it.")

Questionable assumption 6. Emotions are purely subjective experiences. How the brain creates an internal representation of the external world, and translates that representation into thoughts, feelings, and action, is a central open question in neuroscience. We know for sure that there is no little person sitting inside the brain looking at a screen and pulling on joysticks. The only things that have access to the patterns of neuronal activity in the brain are other neurons in the brain. How neurons "decode" the information represented by other groups of neurons and pass that information on to yet further groups of neurons so as to organize and express thoughts, emotions, and actions, is a deep mystery that we are far from solving. This leads to the following scientific questions:

"How exactly do emotions arise in the brain? Can we separate the subjective, conscious experience of emotions from the existence of emotion states per se? Do emotions always have to be conscious? If so, how should we study them in animals, who may or may not be conscious and, in any case, cannot tell us how they feel?"

As we elaborate in the next section, we believe it is critical to distinguish between emotions as internal functional states, and conscious experiences of emotions (often called "feelings"). Emotions and feelings are not the same thing, although they are of course closely related. Most of this book is about emotions, not about feelings. We review some of the work on feelings near the end of this book.

BOX 1.1. The fallacy of the homunculus.

A homunculus, literally "little person," refers to the idea that inside your brain there is a separate observer, something that can watch and interpret the activity of all the other brain regions in the same way that an external scientist might be able to record from your brain and make sense of its processing. The idea of a homunculus has a long history in psychology and the philosophy of mind. It fundamentally arises from a confusion between different levels of description. On the one hand, we know that humans and animals have emotions (and many other mental states). On the other hand, we know that these mental states are produced by the brain. It is therefore tempting to conclude that emotions must literally be found in the brain if we only look with sufficiently microscopic tools.

But producing emotions is not the same as having an emotion. By analogy, there are many places in the brain that participate in producing vision, from the retina to the thalamus, to the cortex. But you cannot find vision in any one of these regions, nor does any of them have the experience of seeing. Or to take one more example: you can drive a car. So who or what does the driving? You can no more be driving by yourself (without a car) than a car can drive by itself (unless perhaps it's a self-driving car). And you can't take apart the car to look for where the "driving" really is located. Driving, vision, and emotion are system properties: they are not properties of any of the constituent parts, but all the parts work together to generate the property.

The most common aspect of emotion where a homunculus fallacy often arises is with respect to the conscious experience of emotion (or, for that matter, the conscious experience of anything else). Unlike the little characters that *Inside Out* put inside the mind of a girl, there are no homunculi in the brain for experiencing your emotions. There are brain systems that make you have a conscious experience of emotion. But the conscious experience of the emotion is a global property of a person (or animal), and the mechanisms whereby it is produced do not themselves have that property.

Toward a Science of Emotion

Without further reflection, it might seem that it should be straightforward to investigate emotions, and to discover how emotions work in the brain. But the assumptions and questions sketched in the first part of this chapter show us that a science of emotion faces some difficult challenges looming ahead. A science of emotion needs to examine most of our initial intuitions about emotions, sharpen vague questions so that they can be experimentally investigated, and confront both empirical and conceptual problems.

Let's take a closer look at one of the major sources of conceptual confusion in emotion science. There is an assumption that different words, concepts, or types of data must refer to distinct things. We will argue instead that one and the same thing can be described with very different words and measured with very different types of data. Consider the thought-provoking image on the next page (figure 1.1), produced by neuroscientist Rebecca Saxe at MIT and published in *Smithsonian Magazine* (December 2015). Saxe got a mother and her infant child to go into an MRI scanner and obtained these images showing their brains.

Saxe writes:

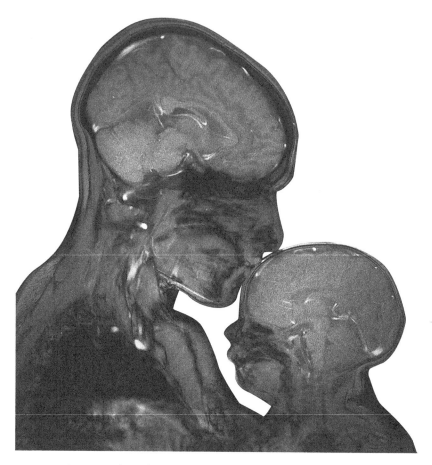

FIGURE 1.1. MRI scan of a mother kissing her child. Reproduced with permission, copyright Rebecca Saxe.

While they lie there, the scanner builds up a picture of what's inside their skulls. Often MR images are made for physicians, to find a tumor or a blocked blood vessel. Scientists also make the images, to study brain function and development. In my lab, at MIT, we use MRI to watch blood flow through the brains of children; we read them stories and observe how their brain activity changes in reaction to the plot. By doing so, we're investigating how children think about other people's thoughts.

To some people, this image was a disturbing reminder of the fragility of human beings. Others were drawn to the way that the two figures, with their clothes and hair and faces invisible, became universal, and could be any human mother and child, at any time or place in history. Still others were simply captivated by how the baby's brain is different from his mother's; it's smaller, smoother and darker—literally, because there's less white matter. Here is a depiction of one of the hardest problems in neuroscience: How will changes in that specific little organ accomplish the unfolding of a whole human mind?

As for me, I saw a very old image made new. The Mother and Child is a powerful symbol of love and innocence, beauty and fertility. Although these maternal values, and the women who embody them, may be venerated, they are usually viewed in opposition to other values: inquiry and intellect, progress and power. But I am a neuroscientist, and I worked to create this image; and I am also the mother in it, curled up inside the tube with my infant son. (http://www.smithsonianmag.com/science-nature/why—captured-MRI-mother-child-180957207/)

As you were reading the above quote, you probably felt a tension between the colder, internal glimpse of two physical bodies shown in the MRI scan and your realization that these are two real people engaged in an affectionate emotional behavior. The MRI scan shows only tissue contrast, revealing bones, fluid, muscle, and brain. At the same time, we know that this is a mother and child—they are people, with thoughts and emotions. Both our everyday view of people and the view made possible with the MRI are of the same thing.

This is perhaps the most critical realization for a science of emotion (indeed, for a science of the mind in general). You can feel emotions. You can infer that other people are having emotions from their behavior. And you can image and record traces of emotions in the brain. These are very different types of data, very different sources of evidence about an emotion. And indeed, they need to be kept separate if one is studying them in their own right—as we shall see, feeling an emotion, having an emotion state, and attributing emotions to another person engage distinct processes in your brain. Nonetheless, your experience of your own emotion, your attribution of an emotion to another person you might see laughing or crying, and the neuroscientist's investigation of an emotion from neurobiological data are not about three different things. They are ultimately all about one and the same thing, an emotion state. You can infer the emotion state in another person from observing their behavior, you can investigate the neural mechanisms of the emotion state through neuroscience experiments, and the emotion state may cause you yourself to have a conscious experience of the emotion. The behavioral observation, neurobiological measurement, and personal experience each can provide evidence for one and the same thing: an emotion state.

To flesh this out a little further, let's view emotion from four different perspectives: the perspective of the behavioral biologist who might be carefully watching the behavior of an animal in the wild or the laboratory (or, for that matter, watching the behavior of a human being); the psychologist concerned with having people talk about and rate their conscious emotional experiences; the psychologist measuring emotional responses in the body, such as changes in heart rate or facial expression (common approaches in the psychology of emotion); and the neurobiologist who is studying (or even manipulating) the function of neurons in the brain (figure 1.2). All four perspectives can be perfectly objective and have an established and agreed-upon methodology—but they are rather different data and often do not use the same language to describe the concepts and methods that relate their data to emotions. Yet all four investigate emotion.

Of those four perspectives, it is especially neuroscience that can show you things you could never get from your everyday knowledge of emotions. What kind of drug will work best for curing depression? Why do some people fear dogs whereas others love them? And, first and foremost, what are the underlying mechanisms that generate emotions— how do neurobiological events in the brain cause tears to run down our face when we are in a state of sadness, and how does this emotion state change much of the rest of our behavior, our attention, our memory, our decision-making? These and many other questions like them are important for treating psychiatric illnesses, for understanding everyday human cognition and behavior, and for understanding the cognition and behavior of other animals. You cannot get at them by just thinking

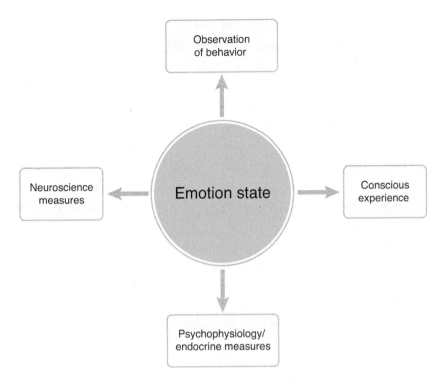

FIGURE 1.2. Emotions can be inferred from several kinds of data. We regularly attribute them to ourselves based on our subjective experience; psychologists might attribute them to us based on our verbal reports of that experience. We also attribute them to other people on the basis of their overt behavior; ethologists might do the same when they observe animal behavior. We might also use additional tools, such as measures of heart rate or blood pressure in the laboratory, to infer that a person is in an emotional state, even when they do not show it in overt behavior. Finally, as neurobiologists, we might look directly into the brain in order to draw conclusions about emotions. All of these measures are parts of a science of emotion.

about your feelings. The aim of a neuroscience of emotion should be to make transparent how and why specific emotions have the features that they do: to explain them through their underlying mechanisms (this is a topic we discuss in detail in chapter 4).

But although this book will focus on neuroscience, our hope is that our broad and functionally based approach will contribute to an integrated science of emotion, a science that investigates emotions through behavior, psychology, and neurobiology. Such a science of emotion should also aim to investigate emotions across species, from worms and insects, to mollusks and fish, to birds and reptiles, to mice and dogs, to monkeys and to people. It would identify specific instances of emotions with particular states of the brain; that is, how they are in fact instantiated in people and animals. But it would also explain why this brain activity instantiates a particular type of emotion in people and animals, and how one might imagine that such an emotion could be instantiated otherwise. Could a nonbiological robot (of the right sort) in principle have emotions? Answering these questions requires a framework that allows us to operationalize emotions in all these different instances. We describe such a framework in chapter 2.

We will speak frequently of "emotion states," or, for shorthand, simply "emotions." But of course, emotions are anything but static states. We refer to them as "states" to keep things simple, but with the realization that they are complex processes that vary in time. We will take apart emotion states and talk about processes in later chapters. But the fact is that we know very little yet about this level of description—so we will keep referring to "emotion states," again, with the full acknowledgment that emotions are in fact temporally extended processes.

Emotions Are Decoupled Reflexes

To see where emotions fit into the picture of complex behavior, let's start by considering a simpler behavior. The knee-jerk reflex, for example, which is a behavior produced when the doctor taps your relaxed knee tendon to cause your leg to move. We call that behavior *simple* for two reasons. First, it is simple because your knee doesn't jerk to pretty much anything except that specific stimulus. It doesn't move if the neurologist shows you a picture of a little hammer. It doesn't move if you just think of a hammer. It also doesn't move if you tap your shoulder. Second, the converse is also true—the tapping of the knee doesn't do much other than move the knee. Your arms don't move, nor do other parts of your body. So the link from stimulus to behavior is very narrow, encapsulated, and modular. In fact, we now know that you can get rid of all of the brain, but as long as the spinal cord is still intact, you will still have the knee-jerk reflex.

Like other reflexes, the knee-jerk is a rigid, narrow, and automatic stimulus-response mapping. You don't feel like you decided to move your knee, you had no urge to move it, you're not thinking or planning to move your knee, and in fact it is hard *not* to move your knee. Such responses have a purpose, but they face a major problem: the world is so complex and changeable that it would take a continuous supplementation of more and more reflexes to deal with all possible kinds of stimuli and adaptive behaviors. You would need yet a different reflex to respond to the ground under your foot, a cup in your hand, food on your lips, and so forth. You would also need some mechanism by which these reflexes could communicate with one another, so that your overall behavior becomes coordinated. Something more flexible is needed for an animal to survive in the world. Enter emotions.

As we will describe in chapter 3, emotions may have evolved out of reflexes, but they show properties that go well beyond what reflexes can accomplish. We may think of the most basic set of these properties as "emotion building blocks," which share some minimal features that distinguish emotions from reflexes (we give a list of these in chapter 3). These features of emotions can be used to infer emotions in many animal species. For instance, flies already show behaviors that meet some of the criteria for emotion building blocks, such as persistence through time (Gibson et al. 2015). There is an evolutionary story to tell about how emotions became more elaborated and eventually took on the set of properties that characterize emotions as we see them in humans. That story needs to be a functional account of what it is that emotions *do*, what function they accomplish in the life of an animal.

An example of such a story can be seen with the emotion of disgust. Disgust evolved so that animals could avoid poisonous or contaminated foods. That might seem like a problem for reflexes to solve: just link the taste of the poison to the behavior of spitting out the food and you are done. Well, it's more complicated than what reflexes can solve. There are many different poisons and contaminants. Some taste bad, some don't. Others might look, feel, or smell contaminated. There are also poisons and contaminants that animals learn about by watching what happens to other individuals (or, in humans, having other people tell you what might happen if you ate them). This kind of learning can be seen in young children for instance—their food preferences change considerably as they mature. Little children often put things in their mouths that we adults find disgusting, but later learn to avoid such foods.

Similarly, even in adults, there can be profound learning of disgust. For example, we may get violently ill from food poisoning and thereafter no longer enjoy that food. Finally, in some situations we may have to overcome our feelings of disgust and ingest something we would otherwise avoid in order to survive: think of a lost explorer who resorts to drinking his own urine in order to avoid death by dehydration (a strategy that only works for a short while, in case you were curious). So there are clearly both innate and learned aspects to disgust, and the disgust state exerts a strong effect on our behavior but can, to a degree, also be overcome. Reflexes just don't have the flexibility to work across these varied cases; they are too narrowly tuned to specific stimuli, and too rigid and uncontrollable in how they cause behaviors.

An alternative is to evolve another layer of control on top of reflexes: to decouple them from their rigid stimulus-response arcs, and instead to incorporate features of them into a different kind of architecture, a central emotion state. We believe that emotions like disgust evolved as central states whose very function is to orchestrate and flexibly regulate such complicated stimulus-response mappings. The emotion state can still be thought of as a functional module similar to a reflex, but, unlike reflexes, that module is portable across a huge range of different situations, many of which we learn about. The scientific study of disgust nowadays views this emotion as having evolved from the need to avoid contamination with pathogens or toxins, which evolved into more complex and social forms of disgust (Rozin 1996) (figure 1.3). But although portable across very different domains, the emotion state of disgust remains concerned with a specific function: the avoidance of passive stimuli that are potentially harmful (the avoidance of actively harmful stimuli might map onto a different emotion, closer to "fear").

The evolutionary invention of emotions as flexible central states resulted in additional benefits. Because emotions need to interface with so many different kinds of stimuli, and stimulus modalities, and because they have to interface with many aspects of motor control, they are also poised to interact with many other central processes. Thus, emotions influence attention, memory, and other cognitive processes—a rapidly growing field in human cognitive neuroscience today (Pessoa 2013). Indeed, it is essential for emotions to interact

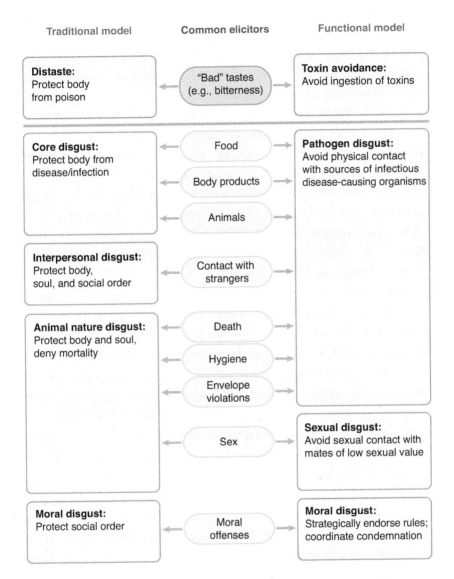

FIGURE 1.3. Functional varieties of disgust. The scheme on the right conceives of disgust as a state evolved for specific functions related to avoidance of potential poisons or contaminants. The different boxes as we go from the top to the bottom of the figure illustrate that disgust has been elaborated from ancestral functions to avoid poison, to more social ones. There is continuing debate about how these functions evolved; this scheme assumes a biological evolution, but Paul Rozin and Jonathan Haidt have argued that cultural evolution may explain some of the functions of disgust. Reproduced with permission from Tybur et al. 2013.

with all of these other processes in order for them to carry out their functional role. For instance, your disgust for a particular food needs to be encoded in memory and needs to redirect your attention. If all you had to work with was the emotion state, without any connection to other cognitive processes, it would bring us back to the encapsulation of reflexes, and their pitfalls.

There are many other properties of emotions that distinguish them from reflexes, a topic we consider in more detail in chapter 3. Of course, reflexes and emotions do not exhaust all of the mechanisms that explain complex behavior. You also deliberate and plan volitional actions, for example. Most of the time, your behavior is produced by a rich and coordinated mixture of many reflexes, emotions, and planned actions. While there is no definitive dividing line, we should thus situate emotions at an intermediate level of complexity in behavioral control, between reflexes and deliberative behavior.

Emotions "decouple" stimuli from responses, as compared to reflexes (Scherer 1994), but they are still fairly automatic compared to deliberative, planned actions. They occupy a particular level of control in the behavior of higher animals: above reflexes, but below planned action. While this helps to situate how emotions fit into the architecture of central states that regulate behavior, it does not yet distinguish emotions from many other central states. For instance, states like motivation, drive, and arousal also fit with much of what we have been saying about emotion. These other kinds of states are considered further in chapter 5. Further delineating emotions requires a more fine-grained look at their properties: what kinds of operating characteristics do we find for prototypical emotion states, and how might this help us to understand the underlying mechanisms, and the functions that they implement? These are topics we take up in the next two chapters.

We have so far sketched only the roughest outline of a science of emotion and what it brings to the table, making three important points. First, emotions need to be studied behaviorally, psychologically, and neurobiologically. A science of emotion needs to be interdisciplinary in the types of methods and data that it considers. Second, emotions function in the flexible control of behavior and are generated by mechanisms within the brain. This generates a list of features of emotions that distinguishes them from reflexes and from several other types of internal states, and that we can test for experimentally (the topic of chapter 3). And third, emotions do not only influence behavior, they also influence many other cognitive processes like attention and memory (and other emotions). Implicit in these three points is that many animals have emotions, not only humans, and, in our view, not only mammals. Everything we have written so far about emotions is applicable to animals and may be investigated in them just as in humans. A science of emotion should thus be interdisciplinary not only in its methods but also in the species it studies.

Questions We Will Not Answer in This Book

In addition to understanding the fundamental principles that are common to all emotion states, there are more detailed questions that a science of emotion will eventually want to address. We will not give answers to these questions in this book, but we will discuss which of them might be answered by empirical investigation, and which might just be semantic issues that are in fact not necessary to answer. Some of these questions return to those motivated by our list of questionable assumptions earlier in this chapter.

1. How many emotions are there? Is the neural representation of a given emotion discrete and separate from that of other emotions, or does it share features or components with other emotions? If the latter, which features of neural representation are specific to a particular emotion, and which are shared? Are some emotions more basic ("primary") than others? We will suggest that this question can be answered by situating emotions in a dimensional space and listing their functional properties. We also believe that many of the words we currently have for different emotions are not going to find a place in a mature science of emotion, which will need to revise the emotion categories that we talk about in everyday life.

2. What is the relationship between the brain representation of an emotion state and that of other related states such as arousal,

motivation, drive, mood, or affect? Are they related (nested or overlapping), or are they different? We will suggest that empirical data indeed show differences between these states, although this question is also partly semantic.

3. What determines the type of emotion that will be evoked by a given stimulus, or whether that stimulus evokes any emotion at all? Are our emotional responses at all "hardwired" by genetics, or are they mainly determined by experience? We will suggest that this question will require further experiments that compare emotions across people and animals, so that we can better understand the full range of emotion states and how they arose in evolution.

4. Are emotion states causative to behavior, or (as William James believed) are they consequences of (reactions to) behavior? Or is it only our *subjective experience* of emotion states that is consequential to behavior? We will argue for the view that emotions cause behavior, not the other way around—but we will also acknowledge that once an emotion has caused a behavior, that behavior can itself trigger further emotion states. So the answer to the original question is: both.

5. How is the "feeling" aspect of an emotion state encoded by the brain, and how does that differ from the "motive" aspect of the state (the part that controls our behavior)? We will argue that feelings are quite distinct from emotions and need to be kept distinct for a science of emotion to make progress. This is a strong view we will return to in the next chapters in some detail to make our argument. However, it is essentially a premise for the rest of the book rather than an answer to a question. We believe that if you accept this premise, there will be benefits for a science of emotion since it becomes more tractable.

6. How will we know when we have fully understood brain mechanisms of emotion? One test would be to be able to predict, with a high degree of precision, how experimentally perturbing the function of any arbitrarily selected brain region or group of neurons would affect any aspect of a given emotion state. We give examples of how this can already be done (in animals) in chapters 5–7. Another test would be our ability to engineer a machine that can

express an emotion state in a manner indistinguishable (by more than superficial visual criteria), from that in a human—that is, to build a robot that has emotions. We think this is possible in principle, even though it is currently science fiction.

What Do We Want to Know about Emotions?

This chapter gave more questions than answers, and discussed examples of what we don't yet know. So what is it that we *do* want to know? We close with a summary of what one might ideally wish for in a science of emotion, and where we focus in this book. Figure 1.4 provides a list of desiderata—things that one would eventually like to have some answer to, even if the answer is that the original question was ill-posed. We do not address all of these desiderata in depth, although we comment on all of them at some point throughout the book.

The first broad desideratum lists some of the most basic phenomena that a science of emotion should explain to the layperson. As you can see from the entries in the figure, this makes a critically important point about the scope of this book. We aim to provide a framework for investigating and explaining emotional behaviors, in both humans and animals. But we bracket the topic of conscious experience, often called feelings. The reasons for this bracketing are entirely practical. We certainly think feelings are important, and we believe that both humans and animals have them. We leave them out in this book because they introduce difficulties—there is no agreement on how to measure feelings, especially in animals, and there is even less agreement on how to investigate them neuroscientifically. A similar bracketing of conscious experience has been taken for many other topics in neuroscience (for example, vision or memory), and yet these sciences have been highly successful. We are wagering that a science of emotion can also get a successful start without needing to tackle conscious experience of emotion at the outset.

The second broad desideratum lists some properties of every good science. Any framework for a science of emotion should provide consistent and objective criteria that produce results that make sense. Let's unpack this important claim. Providing consistent and objective criteria means

Desiderata	This book
Layperson's phenomena to explain	
Conscious experience of emotion in humans	No
Conscious experience of emotion in animals	No
Emotional behavior in humans	Yes
Emotional behavior in animals	Yes
Scientific features	
Objectivity (facts, not opinions)	Yes
Accessibility (connects with other disciplines)	Yes
Clarification of terms (not ambiguous)	Yes
Taxonomy (categories of emotion)	No
Experimental approach	
Modern measurements (sensitive, comprehensive)	Yes
Causal intervention and manipulation	Yes
Broadly applicable methods	Partly
Generalizability	
Multi-level account (abstraction, functional description)	Yes
Comparisons across species	Yes
Comparisons across individuals	Partly
Applications	
Discovery (experimental investigation)	Yes
Assessment (evaluation, assessment, diagnosis)	No
Intervention (treatment, cure)	No
Construction (AI, robots)	No

FIGURE 1.4. Desiderata for a science of emotion. This figure summarizes what aspects of emotion are the focus of this book.

that we agree on how to determine whether there's an emotion or not, and that we do so consistently across cases. Thus, we should have some agreement on what kinds of measures or outcomes would rule in favor of a particular animal being in an emotion state or not. Moreover, we should apply similar criteria across animals and humans.

We also need these criteria to produce results that make sense. Making sense means providing some correspondence with our prescientific intuitions. Of course, there can be disagreements, and of course there are problematic cases. But this doesn't detract from the fact that there are clear cases on which most people agree—if this were not the case, we couldn't be talking about emotion. Most people agree that humans, monkeys, dogs, and cats have emotions. Most people agree that cars and clouds and laptop computers do not have emotions. So whatever account a science of emotion ends up with, it should respect these starting points—if it does not, something probably went wrong with the science. This second broad desideratum, then, asks of a science of emotion that it gives us some objective way to interpret data, such that we could decide whether an organism is in an emotion state or not.

The other entries under "scientific features" provide related properties that help to translate the objective approach into a useful science in practice. If a science of emotion were so encapsulated that it remained within psychology, or even within philosophy, that would be a pity. If psychology and neurobiology used different terms, or terms with different meanings, that would be a bad sign. So we want to aim for a science of emotion that is integrative in the sense of connecting across disciplines. It should provide something meaningful and useful to the philosopher, psychologist, and neurobiologist alike.

The third and fourth broad entries, experimental approach and generalizability, sketch further features that are more specific to our particular framework. One could certainly imagine a science of emotion that does not exhibit all, or even any, of these features—but such a science would not look modern, would have a narrow domain of application, and would fail to capitalize on the most exciting current opportunities in techniques, analyses, and objects of study. This would also be a pity. We would like a science of emotion to use cutting-edge methods (for instance, optogenetics, the ability to experimentally cause an emotion in an animal by using light to manipulate neurons in its brain; see chapter 5). We would like to have computational models and mathematical equations that can allow us to better explain and predict emotions. And we would like to be able to study emotions not only across human subjects but also in apes and monkeys and dogs and cats and rats and even flies and octopuses. In short, the entries under these third and fourth broad desiderata aim to make a science of emotion a modern, vibrant, and frankly exciting and fun science.

Finally, the last entry in our list of desiderata is about applications. These are extensions: what you could eventually do with a science of emotion, once you have it figured out. This book only applies to the very first subentry: we can begin to use a science of emotion to discover facts about emotion. Which animals have emotions? Which brain mechanisms produce emotions? How can we measure emotions? Once we have accrued more data from such a science of emotion, we will be able to apply it to assessment (for instance, diagnosing whether a patient is depressed or not), intervention (treating the depression), and even construction (building robots that have emotions). We think all three are possible to do, but none of them are easy to do yet, and so we do not treat them within the scope of this book.

CHAPTER 2

A Framework for Studying Emotions

Before we discuss the neurobiology of emotion, it is important to provide a foundation for what we mean by "emotion." This chapter will discuss a broadly philosophical view according to which emotions are functional states. Explaining exactly what this means will require a little patience but is very important in order to avoid later confusions. This is especially critical, since there is often little agreement between different scientists in their use of the word *emotion*. Even some neurobiologists, such as Joseph LeDoux at New York University, whose laboratory has conducted some of the most influential studies on fear conditioning in rats, advocate that we no longer use emotion words (like fear), or even the generic word "emotion," for certain domains of scientific inquiry (LeDoux 2012). LeDoux gives some good reasons for his prescription: the fact is that scientists are sloppy in how they use the word "emotion" (LeDoux 2017). In particular, they sometimes mean it to refer to feelings, and they sometimes mean it to refer to behaviors or internal states. It becomes a problem when these different meanings are conflated.

We agree with LeDoux's worries but draw a different conclusion. We firmly believe that the word *emotion* should be used in all scientific investigations of emotion—albeit with clarifications where needed. In addition, we think that some words for specific emotions should be used in scientific discourse—but with more clarifications needed. Likely examples of words that the future emotion scientist will use are fear, anger, and disgust. However, as we noted in chapter 1, we also believe that these terms will need to be revised as a science of emotion progresses. Furthermore, there are many words in our language (and more in other languages, as we noted in the previous chapter) that some people would describe as pertaining to emotions, but which we would feel are premature to use scientifically because we do not yet know whether they really refer to distinct types of emotion states (or emotions at all). That doesn't mean we can't have a science of emotion, it just means that for the time being we should focus on a restricted set of words and concepts. This is also one reason our book is not focused on specific emotions (although we do use them as examples), but rather on emotions generically.

Currently, there are a number of debates in the philosophy and psychology of emotion that are impossible to resolve because they turn on semantic rather than empirically testable questions. It is easier to provide a clear framework for investigating emotion if we focus on emotions in general, rather than getting bogged down in questions about what specific emotions there are, or how fine-grained we want our taxonomy of emotions to be. We need to patiently build up a science of emotion, beginning with those aspects of emotion that are the most amenable to scientific investigation, and then slowly broaden our domain as we gain a better and better purchase on the topic. Whether love, doubt, or awe are specific emotions to investigate scientifically are not questions that we need to answer at this stage.

Warm-Up: Neuroscience Questions about Emotion

We'll warm up toward the conceptual issues of this chapter with a short exercise, similar to what we did in chapter 1 when we listed some assumptions and questions, but more methodologically fleshed out. Let's pick a question that should be of interest to scientists who want to understand emotion and try to triangulate on how to approach that question: what's known, what's not, and how could we find out?

Broad Question: How does the brain generate the state that corresponds to an emotion? What is happening inside my brain when I am in a state of anger, fear, or sadness?

Working hypothesis: Emotions are *internal brain states* that *cause* observable external changes in behavior; observable internal physiological changes in the state of the body; changes in other mental states; and, under some conditions and in some species, changes in what we are consciously aware of ("feelings," which, in humans, can often be verbally described). *Emotion states* are generated by spatially and temporally characteristic patterns of neuronal (electrical) activity in the brain, together with associated changes in brain chemistry (hormones, neuromodulators, and such). They orchestrate a battery of responses. If we understood how emotion states are generated, then in principle we could (a) predict a person's internal emotion state by "reading out" (measuring) their brain activity; (b) induce a specific emotion state by manipulating specific brain circuits or chemistry; and (c) restore normal function to a maladaptive emotion state (for example, depression) by readjusting specific brain circuits or chemistry.

Can't we already do all that? Many neuroscientists would argue, for example, that we can already:

1. Predict whether a person is "feeling afraid" by decoding this from patterns of brain activation;
2. Elicit the experience or display (in animals) of specific emotions (for example, panic) by electrically stimulating specific regions of the brain;
3. Treat depression by administering an SSRI (selective serotonin reuptake inhibitor, a class of antidepressant drugs like Prozac), or in some cases by deep brain stimulation (DBS), a more invasive treatment for depression that electrically stimulates the brain.

So, what's left to understand? Even if we *could* do all those things (and on close examination, there are doubts that we can), they would not be sufficient to answer the question of how an emotion "works" in the brain. The reasons can be seen by analogy:

1. We can predict whether a car is moving or not, and how fast it is moving, by "imaging" its speedometer. That does not mean that we understand how an automobile works. It just means that we've found something we can measure that is strongly correlated with an aspect of its function.

   Just as with the speedometer, imaging activity in the amygdala (or anywhere else in the brain), in the absence of further knowledge, tells us nothing about the causal mechanism and only provides a "marker" that may be correlated with an emotion.

2. We could walk into a modern apartment, and by trial-and-error pressing, determine which button on a control panel turns on the lights in the kitchen. That does not mean that we understand *how* pressing that button leads to turning on the lights in the kitchen (rather than, for example, the living room)—especially if we don't know anything about electricity and electronics.

Just as with the case of turning on the lights, the utility of SSRIs in treating depression, like many medical treatments, was discovered by accident. Physicians routinely use therapeutics whose mechanism of action is not understood, but which usually (but not always) work based on clinical experience. The problem arises when there are depressed patients who do not respond to the treatment—then we are left helpless. Trying to cure these patients without understanding how the brain generates an emotion state would be like trying to cure the bubonic plague in the fifteenth century without understanding that bacteria and viruses cause infectious disease.

The above examples and revealed gaps in our current understanding highlight the general structure of how we would like to be able to understand emotions. A psychologist might look for a somewhat different kind of understanding, and be satisfied with a different kind of explanation, than a neurobiologist. In this book, we want to provide a starting point that would work both for the neuroscientist and for the psychologist. We want some kind of picture of what an explanation of emotion would look like, even in principle, so that we have a concrete goal, and some metric of whether we are making progress toward it. We can apply the kind of dialectical inquiry that we applied to the broad question above to a more specific question to see how this would unpack.

More specific question: What do we understand about the neurobiology of "fear," and what is left to understand? Fear is arguably the best-understood emotion, from the neurobiological standpoint. To illustrate the gaps in our knowledge/understanding, here is a brief and rather caricatured account (necessarily leaving out a lot of details) of one standard view of what happens when we become "afraid," in response to a threat:

1. As a threatening stimulus is detected, activity in our amygdala increases.

2. Increased amygdala activity drives increased activity in "down-stream" targets of amygdala neurons (cells that receive input from the amygdala), which in turn control different aspects of a fear response such as increased freezing (behavioral response), increased heart rate (autonomic response), and increased release of stress hormones (endocrine response). The neural structures mediating these varied effects are located in regions such as the hypothalamus and midbrain periaqueductal gray (PAG), as illustrated in figure 2.1. Not shown in the figure, but equally important, are less well-understood projections from the amygdala to many regions of the cortex, which probably mediate many of the effects than an emotion has on other cognitive processes.

3. We consequently become more aroused, expectant, alert, and sensitive. Our palms sweat, our mouth becomes dry, and there are many other patterns of autonomic and bodily response (box 2.1). We may stop eating or drinking (or whatever else we were doing) and look around. We may defecate or urinate uncontrollably, depending on how afraid we are. We experience a subjective "feeling" of fear. If asked, we would reply that we are afraid or anxious.

4. We may be relatively more or less afraid, depending on the intensity or proximity of the threat, the context in which we experience it (a dark room versus outdoors in bright daylight), and our previous experience with the stimulus. Our physiological and behavioral responses may differ depending on the level of fear, as may our conscious experience (for example, uncontrolled urination may only occur with a very high level of fear).

5. Our decision-making capabilities, and the decisions themselves, may be affected while we are in a fear state.

6. After a while, if no threat materializes or we have experienced no harm, the fear gradually subsides and we return to going about our business.

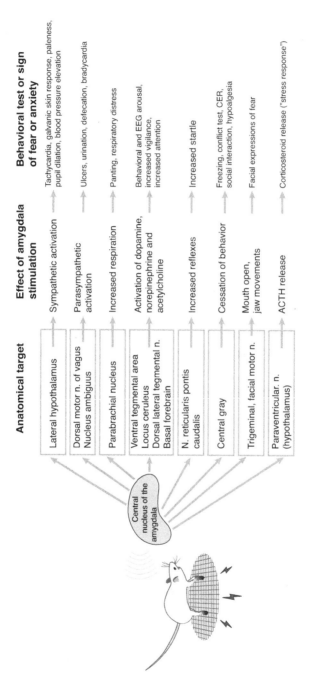

FIGURE 2.1. Orchestration of many components of a fear response by the amygdala. This involves projections from the central nucleus of the amygdala to regions in the brainstem and hypothalamus, based on data obtained primarily from rodents. However, there are many more projections than the ones shown here, including projections to the cortex, and there are also inputs to the amygdala from almost all of the targets to which it projects, making it complicated to disentangle cause and effect within the system. See chapter 6 for details. Modified with permission from Davis 1992.

Anatomical target

	Effect of amygdala stimulation	**Behavioral test or sign of fear or anxiety**
Lateral hypothalamus	Sympathetic activation	Tachycardia, galvanic skin response, paleness, pupil dilation, blood pressure elevation
Dorsal motor n. of vagus Nucleus ambiguus	Parasympathetic activation	Ulcers, urination, defecation, bradycardia
Parabrachial nucleus	Increased respiration	Panting, respiratory distress
Ventral tegmental area Locus ceruleus Dorsal lateral tegmental n. Basal forebrain	Activation of dopamine, norepinephrine and acetylcholine	Behavioral and EEG arousal, increased vigilance, increased attention
N. reticularis pontis caudalis	Increased reflexes	Increased startle
Central gray	Cessation of behavior	Freezing, conflict test, CER, social interaction, hypoalgesia
Trigeminal, facial motor n.	Mouth open, jaw movements	Facial expressions of fear
Paraventricular. n. (hypothalamus)	ACTH release	Corticosteroid release ("stress response")

Central nucleus of the amygdala

BOX 2.1. Can emotions be read out from psychophysiology?

There is a long history of the involvement of the autonomic nervous system in the coordinated effects of an emotion state. The psychologist William James first proposed a specific theory, arguing that autonomic as well as other bodily responses could be sensed by the organism and in turn cause conscious experiences of the emotion. More recent views, discussed in chapter 9, provide a more detailed and neurobiological version of this view. They argue that interoception (our perception of the state of the body) underlies our experience of emotional feelings. These views, together with the ease of measuring a few autonomic variables, have led to the widespread use of psychophysiology in emotion research (Cacioppo, Tassinary, and Berntson 2007).

Psychophysiological measures in the laboratory commonly include measures of skin-conductance response (a measure of sympathetic autonomic arousal), electrocardiogram (measuring your heart beat, from which both sympathetic and parasympathetic measures can be derived), pupil dilation (easily obtained with modern eye trackers, and in part an autonomic measure), respiration, blood pressure, and facial electromyography (sticking small electrodes to the skin of the face to measure tiny movements of the facial muscles and expressions). There are many other measures available in principle, since the autonomic nervous system in fact innervates all bodily organs. Additionally, measures of hormones and metabolites in the blood, urine, or saliva can provide coarse measures of more tonic aspects of emotion, such as arousal due to stress.

There is little question that emotion states cause changes in autonomic nervous system activity, and hence changes in smooth muscle, internal organs, and endocrine measures that are controlled by the autonomic nervous system. However, so far there is no good evidence that the measures typically obtained in the lab, by themselves, are sufficient to discriminate among specific emotions (Siegel et al. in press).

Yet lay people believe that specific emotions are felt in specific parts of the body. A study by the Finnish researcher Lauri Nummenmaa and his colleagues provided a neat visualization of this (Nummenmaa et al. 2014). The researchers designed an online tool through which participants could, with a mouse, fill in those parts of their body where they believed they felt certain emotions. They found some clear patterns (figure 2.2). This same team has also investigated how such body maps of emotions change during development, and how they might differ across cultures, and found interesting differences for both.

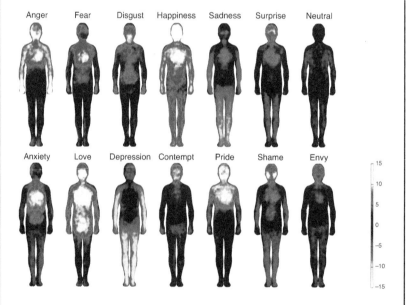

FIGURE 2.2. Bodily maps of people's beliefs about emotional feelings. When asked where they believe certain emotions are felt, Western adult participants tend to indicate the hot regions shown here (red shows an increase, blue shows a decrease, all are relative to neutral). These findings show that people's *concepts* of and *words* for emotional feelings associate them with sensations in specific body regions. Importantly, the findings do not show that actual experiences of emotions are associated with feelings in these body regions, nor that in fact there are specific changes of any sort in these body regions during specific emotion states. The study is thus purely about what people think or believe about emotional feelings. Reproduced with permission from Nummenmaa et al. 2014.

That sounds pretty complete. What is left to understand?

1. We don't understand how individual brain structures contribute to emotions. There is more to the emotion of fear than activity in the amygdala, and there is more to the amygdala than processing fear.

   a. Destruction of the amygdala in animals (and in some rare human patients as well; see box 8.2) impairs some forms of fear but leaves others intact.

   b. Experimental stimulation of some parts of the amygdala is sufficient to evoke fear responses, but so is stimulation of other brain regions that are not directly connected to the amygdala.

   c. The amygdala also contains cells that are activated in response to pleasurable or rewarding stimuli. So just because the amygdala is activated during fear, this doesn't mean that all instances of amygdala activation signify fear. Do the same cells in the amygdala encode both fear and reward, depending on circumstances, or do different types of amygdala neurons participate in these two states? We will see the answer to this question later in this book.

2. We don't understand the causal role that brain structures play in emotion. Even if activation of the amygdala were necessary and sufficient to generate fear, that doesn't mean that the associated central state of "fear" is generated exclusively by the amygdala.

   a. The state of fear may not be generated by *any* single brain structure, but may be generated by a distributed pattern of activity involving many different brain regions. That pattern of activity may be difficult to visualize or measure with current techniques. After all, if you cut out just the amygdala from a brain and put it into a dish, you could stimulate it as much as you like, and it would seem absurd to claim you are causing a state of fear in your dish. Insofar as the amygdala generates fear, it can do so only within a hugely complex network of other brain structures with which it is connected.

   b. Nonetheless, there may be centers or regions in the brain that *orchestrate* fear states, just like the amygdala seems to do in a healthy, connected, brain (see figure 2.1). However, it is un-known whether there are multiple such centers that act inde-pendently, or that control different types or degrees of fear. Coming back to the analogy of hitting switches in a house to turn the lights on and off: we just don't know the wiring dia-gram for fear in the brain.

3. We do not understand how the "level" of fear that we experience is encoded in the brain, or how different responses (freezing, flight, urination) are expressed as a function of different levels of fear.

   a. There is no evidence that a higher level of activity in the amyg-dala simply causes a higher level of fear. It could equally be the case that additional brain structures are recruited at higher levels of fear, or that distinct neuronal populations or distinct patterns of activation in the amygdala encode different degrees of fear.

   b. "Fight-or-flight" responses to fear are not a simple package: different brain circuits control flight, freezing, and fight. The brain has to decide which of those behaviors to engage in at any given moment. We do not understand how the brain performs this decision-making process.

4. We do not understand how a central state of fear affects our decision-making abilities, or other "cognitive" functions; con-versely, we do not understand how our cognitive processes can control our emotions.

   a. For example, adult humans are able, to some degree, to control their emotions (box 3.4). How does this work? There is very little mechanistic understanding of this process from controlled experiments in animals, even though it is a huge topic in human psychology.

   b. More broadly, an emotion state, like fear, cannot cause behavior just by itself. Neither can an isolated memory, state of attention, or perception. In the economy of different types of states of the mind and brain, many different kinds of mental states need to interact in order to produce behavior (or conscious experi-ence). We don't remotely understand how this works. A full understanding of how emotions cause behavior will ultimately require a full understanding of how all other types of psycho-logical states contribute.

The above "warm-up" example may seem to imply that we know almost nothing about how the brain produces emotions. How can we make progress, given so many things we don't know? One of the most important overall take-home points of this book is that a science of emotion will need to tackle this challenge in two ways.

First, *zoom in*. Find a specific, focused question, design a clear experiment with the best methods, and find an answer. Your answer will be a small answer; you won't suddenly understand all of emotion. Instead, your answer will correspond to one of the many subquestions we listed above and be one piece in the very large puzzle of understanding how emotions interact with many other states to produce behaviors. For this to work, your experiment needs to be part of a larger scientific enterprise. A science of emotion must be cumulative, one in which comparisons can be made across results obtained with different methods, and from different species.

Second, *zoom out*. You will need to go from the implementation of an emotion in the brain to a description at a more abstract level. That is, you will need to attempt to understand the emotion as a functional, or computational, process. You will need to be able to interpret and situate the results from your experiment in some way. To answer the question, "what is an emotion," you will need to answer what it is that emotions *do*: what *causes* emotions and what in turn is *caused* by an emotion state? The rest of this chapter is devoted to unpacking what this means.

Toward a Functional Definition of Emotion

Although we have been emphasizing the brain in what we have written so far, emotions should not be defined, fundamentally, in terms of neurobiology. This statement may come as a surprise, but in fact we believe that it applies not just to emotions but to all psychological states; they are states of the brain, but they need to be defined at a more abstract, functional level. For if emotions were defined literally as brain states, we would have to provide a different definition for humans, for flies, for octopuses—because these species have completely different kinds of nervous systems. We therefore begin by stating that emotions *happen to be* states of the brain—they can be implemented in the brain—but that a more abstract kind of description is needed to provide a useful and broad definition. We also need information about stimuli and behaviors that, together with information about the brain state, tell a causal, functional story, a story about how stimuli cause brain states that cause behaviors (figure 2.3).

To understand what an emotion is, we have to start with a functional definition of emotions. We may not yet have full functional definitions of specific emotions, but functional accounts are the *type* of description we need in order to judge whether a state is an emotion or not, and, if it is, in order to judge what *specific* emotion is present.

A functional account identifies the state by its causal relations (what does it do?). It does not identify the state by how it is constituted (what is it made of?). One way to think of this is by reference to the boxes in figure 2.2. If you had very different inputs going into the emotion state, and had it hooked up to very different outputs, it would no longer be an emotion state. Or think back to the amygdala in a dish that we mentioned earlier in our "warm-up" example: yes, the activation of the amygdala can produce some types of fear in a healthy brain that is connected with the amygdala. But stimulating an isolated amygdala in a dish does not produce fear, because we have removed the functional (causal) role normally played by the amygdala. Functional definitions describe causal effects in an abstract manner that doesn't depend on the physical substance that is implementing a state.

We actually have lots of functional definitions we encounter in everyday life. Consider a clock. If I asked you to tell me what a clock is, you would hopefully not begin by saying, "well, it has to have a pendulum and gears inside," or, except as a joke, "a clock is something that weighs 1.2 pounds." Definitions like that would immediately exclude most clocks, including my computer's clock and atomic clocks. So what's a clock? It's a device that measures time. That's a functional definition.

Functional definitions are commonplace in cognitive science and philosophy of mind. There is a well-known position, psycho-functionalism,

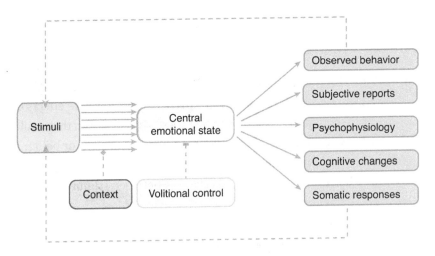

FIGURE 2.3. Emotions are functional states. A very minimal architecture for emotion states is shown in this figure to help orient the discussion in the text. Emotions are typically caused by certain events (stimuli), and in turn cause other events (like behaviors). Therefore either stimuli or behaviors can provide some partial evidence for a central emotion state. The same applies to verbal reports of subjective experience—it is one of many possible sources of evidence for a central emotion state. Notably, many different stimuli can cause the same type of emotion state, and that emotion state orchestrates many different behaviors, showing a "fan-in," "fan-out" architecture (the multiple arrows going into, and out of, the emotion box in the figure). This property of emotion is called "generalization" in figure 3.2.

that proposes that all mental states, whether emotions or any other mental state, are functional states in this sense. They are defined by their causal relations to sensory inputs, behavioral outputs, and also causal relations among one another. In the case of an emotion, this view would say that emotions are functional states that are typically caused by sensory inputs, that typically cause behavioral outputs, and that also cause changes, and can be caused by, other mental states like perceptions, memories, attention, and so forth (for more details, see Adolphs and Andler 2018).

You will note that there is a large promissory note here: the phrase "and so forth." The fact of the matter is that we do not yet have a full scientific inventory of all mental states, and so we cannot yet give a full functional definition of an emotion state (or any other mental state). Nonetheless, we can provide a partial account that, we believe, is enough to point us in the right direction (and will help, bit by bit, eventually to construct a full science of the mind).

We think that emotions should be functionally specified in the above sense if we want to investigate them neuroscientifically. We try to do this by experimentally controlling stimuli and measuring behavior, and by trying to hold constant (or measuring as covariates) other cognitive states. Thus, I could show you a scary predator and measure effects of the sort schematized in figure 2.4 to begin a study of fear. But this would not work if you suddenly change your memory, attention, or belief; if you cannot recognize the stimulus, do not attend to it, or believe it is no longer harmful, the functional relations between stimuli, emotion state, and behavior also change.

This operational emphasis on stimuli and behavior requires us to also have a functional account at a broader scale, a functional account that explains why an emotion is adaptive, and how it evolved, relative to the environment in which a particular organism lives and the ecological challenges it has to solve. After all, it is only such functional effects that evolution "sees." Anything that evolved by natural selection must have done so in virtue of what it does, what causal effect it ultimately has on how well an organism is able to function in its given environment. Unlike the case of the clock, where we built it and thus can authoritatively say that its function is to keep time, the adaptive functions of emotions require detective work to ferret out. In fact, the majority of Darwin's highly regarded monograph, *The Expression of the Emotions in Man and Animals* (Darwin 1872/1965), is focused on deducing the adaptive function of different expressions of emotions—like blushing, for example (whether he was correct about these is another matter).

The broad evolutionary explanation, and the functional explanation in terms of relations to stimuli and behaviors, are just two of the scales at which one can create functional descriptions of emotion. Many more are possible. We could give a functional account of the amygdala's role by explaining what it contributes to a distributed neural network—the computations it carries out, the way it transforms the inputs that it receives from various brain regions into outputs that it sends to other brain regions, and how this contributes to particular emotion states. We could even give functional accounts of single neurons, although the functions that single neurons implement are of course going to be elementary; they are not going to have anything obviously to do with an emotion state at all, but instead are going to correspond to the very many low-level building blocks out of which more complex functions can be constructed. At the other end of the spectrum is the functional explanation of emotion at the level of the entire organism and how it interacts with its environment. That evolutionary point of view is the level that ultimately anchors what the emotion is really about, in the same way that saying that clocks keep time is ultimately what explains what an entire clock does (even though, similarly, we can give functional explanations of the internal workings of any particular type of clock).

People often talk as if they have a nonfunctional concept of emotions—saying that "fear is in the amygdala," or that there are specific neurons "for fear," would be such examples (examples that, as we noted earlier, are seriously wrong). Perhaps our ordinary (nonscientific) concept of an emotion is one that is not functionally based, or perhaps it is. Here, we want to know what kind of account of emotions a *scientist* should strive for in order to anchor an investigation of emotion. What conceptual foundation should a science of emotion adopt in order to have an objective topic of study? What foundation should it adopt in order to produce an integrative, cumulative science that can work across ethology, biology, psychology, and neuroscience? Such foundations are essential in order to eventually give us a mechanistic explanation of emotions, and to tell us how emotions fit into our picture of the rest of the brain and mind.

Proper Functions and Malfunctions

You might think that what we just said about emotions being functionally defined is incompatible with them being brain states. Didn't we just give the example of clocks? Those, we saw, were defined as timekeeping devices, not necessarily as gears and pulleys, or transistors, or sand seeping through an hourglass—all very different physical realizations of clocks. Analogously, if emotions are functionally defined, then is it wrong to think of them as being brain states?

Above we argued for a functional definition so that our concept of emotion states could apply to many different species of animals (and even robots, in principle). But whether or not the functions of emotions, once we have them figured out in sufficient detail, can actually be realized in all kinds of nervous systems, or in systems other than brains, is an empirical question. Maybe they can; maybe they cannot. Not just any system could be a useful clock, and so too, not just any system can instantiate an emotion—perhaps only certain kinds of brains can.

When you have a particular emotion in a particular instance, that corresponds to a pattern of activity in your brain (as we will see in subsequent chapters, probably a very complex and distributed pattern). When I am happy right now because writing this book seems to be going well, that emotion is a brain state. When I wake up anxious because I remember how much I still have to write, that is also a brain state, and a different brain state.

But this is also where it gets complicated. I may be afraid on many different occasions, and so may you. Those are all instances of fear. Are they all the same brain state? Probably not exactly the same, especially if we consider two different people, which means brains with distinct neural network composition. So then the question is, how similar are the instances of fear neurobiologically, and what would you use to measure that similarity?

How do I know that my state of fear doesn't look more like your state of anger in the brain? How do I compare the state that my brain is in during fear to the state of an animal's brain during fear? That is, even though any given instance of an emotion is a brain state (this is what philosophers call "token identity"), it is much less clear whether a particular *type* of emotion (say, fear) corresponds to a *type* of brain state (Measured how? Amygdala activation? Certain neurotransmitter levels?).

This problem of "type identity" is well known in the philosophy of mind in general, and we fully acknowledge it here also. Returning to our example of the clocks, we could all agree that clocks are in fact physical devices. But if you just showed me an hourglass with sand, a computer chip, a sundial out in the desert, and a grandfather clock, just by looking at those physical devices, it might be impossible to figure out what they have in common. I might be inclined to put the hourglass into a store that sells antique glassware, the computer clock with radios, and so forth. Unless I know their function as *clocks*, their physical characteristics don't seem to provide me with good clues for how to group them together into a category. We'd need to observe them in action to get some clues. Similarly, we believe it is absolutely indispensable to have animal (and human) *behavior*, ideally in the natural environment, as one source of data; that is, a science of emotion, among other things, needs ethology. Without behavior, you can never figure out what emotions actually do, and so can never figure out what type of emotion a brain mechanism might be implementing.

How to assign physical states (like brain states) to functional categories (like specific emotions) is an empirical issue: it requires observation, experiment, and data. Even though we have just stressed the indispensability of behavioral data, the neurobiological data can also help to constrain and understand the functional explanations that we seek, a major reason for wanting to do the neuroscience. The reason is that we don't know the detailed functional criteria for different types of emotions except in the vaguest terms, and so information about the neurobiology will surely help to constrain our functional characterization.

Maybe it will turn out that different animal brains, or even a single human brain, can implement emotions in many different ways—maybe there are multiple mechanisms that have evolved to subserve one and the same (or a related set of very similar) functions. Maybe it will turn out that there is only one way that brains can implement emotions, if only we have a description of the neurobiology at the right level of abstraction (maybe a particular set of computations needs to be performed, and that can only be done by a particular circuit architecture, for example).

Probably, the truth lies somewhere in between these extremes. We think it likely that evolution invented several different detailed mechanisms for implementing a shared set of computational functions that work together to coordinate an emotion state. Neurobiology, put together with ecological investigations of behavior, would tell us what those mechanisms are, what computational similarities they share, and how they can implement emotions in a particular animal species operating in a particular environment.

By analogy, we can return to the clock example. My grandfather clock at home is a clock because it keeps time. That is its function. A particular time, say three o'clock in the afternoon for my clock, is identical to a particular physical state of levers and pulleys in the clock. The two are not incompatible. A particular time given by the clock just is a physical state of the clock, even though the time is defined functionally.

Yet there is *some* flexibility in how function maps to implementation, even for a specific single device like my grandfather clock. I could also use the clock to tell me the time in Japan, if I like; I would just always subtract eight hours. Now I am using the clock still to keep time, but for a slightly different function: the time in Japan, rather than in California. The physical state of the clock hasn't changed, but its functional state has (slightly).

I could also take the clock and hang it in the room in which our cats sleep so that the steady ticktock lulls them to sleep. Or I could put the clock on a stack of papers to hold them down by its weight. We would not conclude from this that the function of clocks extends to inducing sleep or being a paperweight. I could use the clock for an unlimited number of silly things, but the very fact that they strike you as silly shows that these take the physical properties of the clock out of context; in a sense, they are functional mistakes.

So there are limits; the physical state of the clock under typical conditions realizes timekeeping, and there may be some flexibility about the functional nature of that timekeeping (which time zone it refers to, for instance, or in the case of my clock, which has no way to distinguish a.m. and p.m., whether it is morning or afternoon). But there are limits beyond which you have simply changed the topic. The same physical state of the clock may make the cats fall asleep, and the same physical state of a brain may give off a certain amount of caloric heat—but inducing sleep and warming up the room are not answers to questions about what clocks do or what brains do.

In the case of clocks, humans intentionally designed the clock, whereas in the case of emotions, the design arose through a long history of natural selection, but both cases have what the philosopher Ruth Millikan dubbed a "proper function" (Millikan 1989). So there is an answer to the question, "what is an emotion," and to give that answer we need to investigate how brains carry out certain functions. There may not be a single answer to the question, but the answer is not arbitrary. Once we have defined constraints on the answer, at least to some extent, it becomes possible to identify whether an answer is in error.

Just like my clock holding down a stack of papers as a paperweight is a malfunction of sorts (the clock is "doing the wrong thing"—something it was not designed to do), so too an emotion of fear in a patient with post-traumatic stress disorder can be a malfunction. Fear didn't evolve to cause PTSD (box 2.2), but the mechanisms whereby PTSD occurs can tell us something about the proper function of fear and anxiety states.

Evolution selected brain states that are emotions, directly or indirectly, so that they could provide adaptive causal functions that pertain to an organism interacting with its environment. Unlike with clocks, we cannot ask a Creator what His/Her intention was in building brains that realize emotions. We need to do a lot of very difficult detective work as scientists to figure out what might have happened during evolution. As with the examples given, there is some flexibility about the details, but there are also constraints.

The task of the emotion scientist is to figure out these details and constraints. How did brain networks get sculpted through evolution so that they can cause adaptive behavior? What processing characteristics are needed to fit the bill for emotions? For example, do you need to focus on particular sensory modalities, or particular kinds of behavioral control? Do you need to process it rapidly or slowly? Can we come up with a list of criteria, as a set of hypotheses, and then investigate the brain to find mechanisms with those features? Chapter 3 begins to sketch such a list, and the subsequent chapters describe how one would go about finding support for the entries in the list (and adding new entries) by studying the brain.

BOX 2.2. Adaptive and maladaptive fear.

Trait-like characteristics of emotions, such as the sensitivity to induce them and the ability to regulate them, exhibit large individual differences across people, with a strong genetic basis. Some people are more easily scared than others; some stay afraid or sad longer than others do. The very flexibility of emotion states, and in particular the decoupling between an immediate stimulus trigger and the emotion state, makes emotions prone to dysregulation: anxiety disorders, for example, constitute one of the most common psychiatric illnesses (all in all, close to 20 percent of the population suffers from an anxiety disorder of some kind in any given year (LeDoux 2015). Animals, too, suffer from anxiety disorders, and many of the same drugs that are used to treat these disorders in humans are also prescribed for pets (Prozac, for instance, works both in people and in dogs who are anxious).

It may appear puzzling to explain anxiety disorders. After all, isn't fear supposed to be an adaptive emotional state? What's adaptive about the exaggerated anxiety responses in somebody who has post-traumatic stress disorder? To understand this, we have to realize that the adaptation is relative to one context (e.g., immediate response to an explosion in order to save your life in the battlefield) that is now exhibited in a different context (the safe home environment). An evolutionary explanation would be that, in our ancestral environment, there were many dangers we no longer face in modern times, and that a distribution of fear responses resulted as quantitative traits that would variably predispose people to become anxious or afraid.

The consequence would be fear responses that can be thought of as false positives from a signal detection perspective. The threshold for detecting fear is too low, and so many stimuli that have a very low probability of being dangerous are misinterpreted as dangerous (Adolphs 2013). This situation makes sense, because the false negatives (e.g., failing to detect a lion that might attack you) could result in death, whereas the false positives (e.g., incorrectly mistaking a tree shadow for a lion) do not have any immediately life-threatening consequence. It is only when false positives cumulatively begin to impair daily functioning, or when their number increases as environmental circumstances change, that pathology becomes evident.

Negatively valenced emotions are often the focus in theories of psychopathology. But in fact, dysfunction can arise from inappropriate engagement of either negatively or positively valenced emotions. Some evolutionary accounts are trying to address such a broader view (Nesse and Ellsworth 2009), in principle making animal models indispensable for understanding all aspects of human emotion dysfunction.

Emotions and Consciousness

The conscious experiences of emotions, or feelings, are typically the defining feature of an emotion for the layperson and the psychologist alike. We haven't said anything about them so far in our discussion of emotions as functional states—indeed, we've left them out. We think this is a good thing. We should not start our scientific investigation of emotion with feelings, because that immediately requires us to provide an account of consciousness—and if you think emotions are hard to figure out, consciousness is *really* hard to figure out! It also leads to concerns about how to study emotions in animals, and even how to study emotions across people in different cultures who may describe their feelings quite differently.

A functional account of what emotions do provides an objective starting point that can be used across many different disciplines that study emotion. It can be used by the ecologist studying emotions in apes or elephants or dogs. It can be used by the neurobiologist wanting to study how mechanisms in the brain can realize the functional criteria of emotions. And it can be used by the psychologist who is interested in how emotions interact with perception, memory, attention, and decision-making. Indeed, it can eventually be used also by the neuroscientist or psychologist who wants to understand how emotion states cause conscious experiences of emotions.

Our emphasis on a functional account is not intended to marginalize the importance of our experiences of emotions. Our hope would be that eventually a neurobiological investigation of emotion would also give us an explanation of how emotions can be consciously experienced. But the nature of our experiences of emotions, and the words we used to describe such experiences, are, in our view, unlikely to be the best guide to identify functional features of emotions.

By way of comparison, consider vision. Our conscious experience of the visual world is similarly fallible. We think we are conscious of everything that is clearly in our visual field, yet experiments that ask people to report on anything they think is in their conscious visual experience put that in doubt. The subjects typically have to direct their gaze and attention onto something specific in their visual field to tell you anything about it. Visual objects are not "in the mind" until we look at them, even though it seems to us as though we are conscious of everything in front of us at once (Noe 2004). Consciousness presents the world to us in a particular way, but what we believe about our visual experience is not the best, and certainly not an infallible, guide to what actually happens in the mind and brain during vision.

One clear example of the difficulties that arise if we equate emotions with feelings comes from the work of the neuroscientist Joseph LeDoux, to which we already alluded (LeDoux 2012; 2017). How do we really know that a cat or dog, let alone a rat or fly, actually feels emotions, or feels anything for that matter? Doesn't the experience of feeling require consciousness? If so, then in attributing emotions to animals aren't we making the much stronger attribution of consciousness? LeDoux is right: it is problematic to attribute conscious experiences of emotions to animals. For that matter, it is problematic to attribute them to other people. But it does not follow from this that we cannot study emotions in animals. It just means that we need to separate the study of emotions from the study of the *conscious experience* of emotions.

The idea to decouple our scientific usage of the word *emotion* from the conscious experience of emotion is not a novel one (Berridge 2003; Damasio 1995; LeDoux 1994; Dolan 2002). We do that with many other words that refer to mental and brain states. For instance, ask the person on the street what they mean by "vision." They will tell you that it is the conscious experience of seeing. But the vision scientist (who could be a psychologist or a neurobiologist—or even a computer scientist or engineer these days) doesn't assume this. There is a lot of work on vision that uses only behavioral measures, a lot of work on vision in flies or in horseshoe crabs (and, indeed, machine vision in computers). These are all studies of vision, but not studies of visual consciousness. There are also people studying visual consciousness, but they say they are studying consciousness, not vision (Koch 2004). We need to make the same move in the scientific study of emotion.

Studies have found fascinating dissociations between visual guidance of behavior and visual conscious experiences. In the phenomenon of "blindsight," for example, patients with rare damage to visual cortices can detect and discriminate visual stimuli even though they subjectively report seeing nothing (they have no conscious visual experience). In fact, one patient was able to walk without bumping into objects, even though he felt blind! (https://www.youtube.com/watch?v=nFJvXNGJsws) (De Gelder et al. 2008). Conscious and nonconscious vision have been studied not only in people but also in animals; monkeys with blindsight have been demonstrated as well (Cowey and Stoerig 1995).

Emotion is no different from these examples (see box 8.2). Although our common usage of this word, and of words for specific emotions, assumes conscious experience, a science of emotion need not, and should not, make the same assumption. A science of emotion should, in the first instance, use behavior, cognition, and neurobiology in its vocabulary. It should not be based on self-reports of feelings in people.

This does not mean we advocate getting rid of feelings and self-reports. We have no doubt that most humans and most higher animals have conscious experiences of emotions when they have emotions. But perhaps some animals (or humans under some circumstances) can also have emotions of which they are not, or cannot be, conscious. Some researchers have claimed that there is evidence for nonconscious emotions in humans (Winkielman and Berridge 2004) (see box 10.1). These are, again, empirical questions. They should not be answered by semantic or conceptual analyses, and the answers should not be assumed before we do the science. Maybe some animals consciously experience emotions, and some do not. The place to begin is with an operationalization of emotion that allows the ethologist observing animal behavior, the psychologist interested in the mind, and the neurobiologist recording from the brain, all to study emotion without needing to study consciousness.

An Experimental Example

What we have written so far makes some specific recommendations about how a science of emotion could move forward, in particular that we need to create more articulated functional characterizations of emotions. What exactly is fear? What is anger? How do they differ functionally—both seem to be about threats, both often show similar behaviors.

Making these distinctions in a functional framework forces us to ask what environmental challenge an organism faces, and how it solves the challenge. What kinds of cognition and behavior need to be coordinated for certain types of environmental challenges? How has evolution come up with a set of common solutions that can be applied flexibly across a range of challenges that all share common features? These functional questions can benefit tremendously—indeed require—neurobiological data. But how do we define "common features" of an environmental challenge? What might a charging tiger and diving hawk have in common that they could both cause fear, for example? More specifically, what might they have in common that a nervous system could evolve to detect?

Neuroscience experiments are beginning to provide initial answers to questions like these. For instance, our Caltech colleague Markus Meister has shown that to elicit specific defensive behaviors in mice, it is sufficient to produce overhead looming shadows, like a hawk would make (Yilmaz and Meister 2013) (figure 2.5). The shadow seems to be an environmental feature common to many threats, and the larger the looming shadow, the more defensive behavior the mouse shows. This experiment maps onto two functional features of threat: threat imminence, the proximity of a threat (figure 2.4), and escape potential, the presence of a hiding place to which the mouse can flee. The next step is now to identify the circuitry in the retina that would detect simple features like such a looming shadow. Researchers have undertaken a similar investigation for olfactory circuitry that might identify the odor of a cat. Ditto for the sound of an approaching predator. These are all sensory cues that the brain can use to detect functional properties of a threat.

But are these defensive responses to a simulated predator accompanied by a central motive state of "fear"? Once we know something about the sets of features that sensory systems can encode as cues to threat, we can begin to ask this question by investigating aspects of brain activity that cannot be explained simply as sensory processing or as motor programs. Telling that story in detail will force us to consider a lot of complexities along the way: how are context, memory, and expectation incorporated into how a stimulus is processed? How can we have partial control over our emotions? And how can emotions arise out of the blue without any apparent stimuli? We believe that the complexities can only be understood once we understand the simplest cases.

**BOX 2.3. The evolution of emotional function:
From survival to social communication.**

What exactly are the functions of emotions? Charles Darwin thought emotional expression evolved as "serviceable associated habits," that is, there was an ancestral survival-related function for the behaviors we now see in the expression of an emotion, but the emotion may have lost that original function (Darwin 1872/1965). How else to explain why we weep tears when we are sad, or make particular facial expressions when we feel certain emotions? It seems hard to explain what good that does, what its function is, in terms of immediate environmental survival.

For example, a fear grimace may serve to widen the eyes in some animals, so that they can see a predator in the periphery—there is experimental evidence for this (Susskind et al. 2008)—and to bare the teeth for possible defense and biting. That may have once been true in humans, too, but when one of the students in my lab makes a fear face upon learning they may have failed a final exam, they are, I hope, not about to bite me. Instead, the facial expression has taken on a different function over evolutionary time; in particular, it now serves a primary function in social communication. In humans, many—perhaps most—emotional behaviors serve a social communicative function in addition to the ancestral functions that are more apparent in other animals.

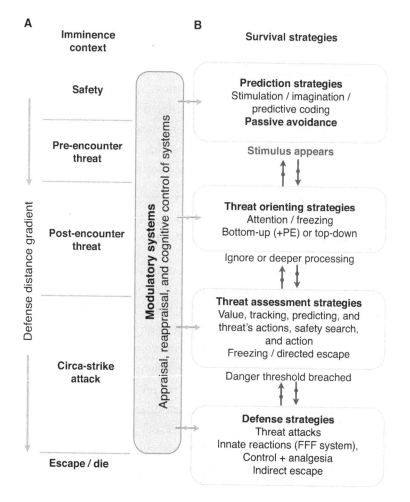

A
Imminence context

Safety

Pre-encounter threat

Post-encounter threat

Defense distance gradient

Circa-strike attack

Escape / die

Modulatory systems

Appraisal, reappraisal, and cognitive control of systems

B
Survival strategies

Prediction strategies
Stimulation / imagination / predictive coding
Passive avoidance

Stimulus appears

Threat orienting strategies
Attention / freezing
Bottom-up (+PE) or top-down

Ignore or deeper processing

Threat assessment strategies
Value, tracking, predicting, and threat's actions, safety search, and action
Freezing / directed escape

Danger threshold breached

Defense strategies
Threat attacks
Innate reactions (FFF system),
Control + analgesia
Indirect escape

FIGURE 2.4. A detailed functional architecture for threat processing. A. On the far left is a dimension that maps the imminence of threat: how close or far away a predator might be to you. B. To the right are functional modules that schematize the kinds of strategies an organism needs to take in order to deal with that threat—from avoidance and orienting for threats that are far away, to defense when they are imminent. Although the paper from which this figure is taken, and the researchers who wrote it, focus on neuroscience studies, this schematic in fact lists no brain structures at all—it could be a blueprint for building a robot that copes effectively with environmental threats. Modified with permission from Mobbs et al. 2015.

One major attempt to provide functional accounts for specific emotions comes from a domain of emotion theory in psychology called appraisal theory. For instance, the psychologist Richard Lazarus provided a fairly detailed list of functional roles that specific emotions served—his so-called core relational themes (Lazarus 1991). Modern appraisal theorists, such as the Swiss psychologist Klaus Scherer, provide similar accounts, only more detailed (Scherer 1988). These accounts provide a sequence of decisions that the brain is hypothesized to make in order to evaluate a stimulus or situation in terms of its relevance to the person and the goals and capabilities of the person.

There is much to like about these kinds of appraisal accounts in the abstract, but they have historically been limited in two main ways: first, they are focused only on humans, and on the experience of emotion; second, although often plausible, their functional criteria for emotions are derived primarily from the intuition of the researchers rather than objective data. Nonetheless, we believe they showcase an approach that will also be helpful for modern neurobiological accounts of emotion—and some neurobiologists are providing accounts that look remarkably similar in spirit to appraisal theories.

For instance, the neuroscientist Dean Mobbs has developed a detailed functional account for thinking about states like fear (Mobbs et al. 2015) (figure 2.4). In this scheme, specific functional properties are mapped along a dimension of threat imminence— how close a predator is to an animal. If the predator is far away, there is only vigilance and monitoring; as it gets closer, specific types of fear states are evoked, going from anxiety to fear to panic. The dimension of threat imminence, together with other parameters (such as the presence of a hiding place and other contextual factors) can all be thought of as inputs to a complex decision-making system that ultimately decides the animal's course of action (freeze and try not to be seen; run away; escape to a hiding place, stand your ground and defend yourself).

From figure 2.5, we can also see how, once we know a bit about the mechanisms behind an emotion, we can understand how emotions might be caused "in error." In the mouse experiment shown in the figure, the hapless mouse freezes in fear when an expanding dark circle is shown on a computer monitor. This happens not to be particularly adaptive in the experiment in the lab, and is in that sense an error. The circuitry did not evolve to detect expanding circles on a computer monitor, but it evolved to detect looming dark shadows swooping down—such as would be reliable threat signals of a bird of prey homing in on our poor mouse. Once we know how the neurobiological mechanism works, we can begin to intervene by directly manipulating specific processing steps in the brain. Such experiments are now possible using modern neurobiological techniques in animals that will also help us understand how we could intervene in the case of humans (see chapters 5–7). It is only through knowledge of this sort that we would have an intelligent approach to treating "malfunctions" of emotion such as depression and anxiety.

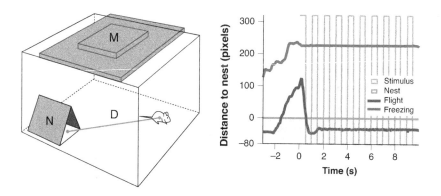

FIGURE 2.5. Response of a mouse to a looming stimulus. *Left*: In the experimental setup, a mouse is in a simple environment that permits the experimenter to manipulate a few cues. One of these is a looming threat stimulus, a dark expanding circle overhead, shown on a monitor (M). Another is the presence or absence of a nest (N), and its distance (D) from the mouse. *Right*: Actual data. A mouse begins to explore the environment, moving around relative to the nest (squiggly blue and red lines at times < 0). At time = 0 the looming stimulus comes on repeatedly (gray vertical bars), causing the mouse either to freeze if it is too far from the nest (flat blue line), or else to race into the nest (green) for safety if it is close enough (red line). Through precise experimental control over stimuli and environment in studies like this, we can begin to understand in detail how the mouse's brain senses visual threat cues and uses them to make an adaptive behavioral decision. Reproduced with permission from Yilmaz and Meister 2013.

Summary

- We discussed a functional approach to understanding emotions: one that explains emotions by their causal effects. Emotions are defined by what they do, rather than by how they are implemented. This is also the level at which evolution has selected them—they carry out specific functions that contribute to our survival.

- A functional conception of emotions also suggests criteria for understanding psychiatric disorders. Sometimes an emotion is not adaptive, and an emotional behavior is a "malfunction" of sorts. Mood disorders, such as depression and anxiety, and PTSD, may be malfunctions like this (box 2.2).

- Feelings (the conscious experiences of emotions) should not be conflated with emotions. We illustrated analogies from the science of vision where commonly used words are also used in a more specific way scientifically—and distinguished from conscious experiences.

- What we need next is a general list of criteria, of functional properties, that emotions might exhibit. We turn to this task in the next chapter.

CHAPTER 3

Building Blocks and Features of Emotions

Emotions are usually considered states of a person. We say that somebody is in a state of fear as they are being chased by a bear, for instance. But we also apply emotions to systems larger than a person. For instance, we can say that "America was in a state of fear" after the 9/11 terrorist attacks. In this case we might be referring to all or most of the people in America, or to America as a more abstract entity in terms of how the country reacted in news, policy, and so forth. The attribution of a fear state to a single person, or to larger social entity, can both be legitimate, if they adopt the functional perspective that we discussed in the previous chapter. In both examples, there is a situation of threat and various functional consequences of an emotion state that collectively attempt to deal with the threat.

As neuroscientists, we are interested primarily in how the functional architecture of an emotion is actually implemented in the nervous system of an individual person, and in turn what clues about the functional architecture the neural mechanisms can give us. We would certainly expect that neuroscience investigations can shed light on "states of fear" that apply to groups of people, just as disciplines like decision neuroscience have helped shed light on collective behavior in economic markets. So, while the topic of our book is the brain, we would expect that topic to be relevant for understanding the behavior not only of single organisms but also of groups and societies that are made up of interacting organisms.

What, then, are the boundaries of an emotion state within a person or animal? One could take the view that an emotion state comprises everything in the nervous system (and perhaps also the body) that is happening when you are in that emotion state. We believe that this is too broad. Not all the events in your brain that are contemporaneous with a state of fear are part of the fear state, only a subset are (figure 3.1).

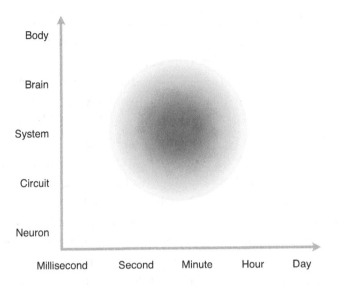

FIGURE 3.1 Spatiotemporal extent of an emotion state. In this book, we treat emotion states as implemented in a subset of the nervous system (less than all of the brain, more than a neuron or circuit), and in events that unfold over the course of seconds to minutes.

The same is true of any other state that psychologists or neuroscientists study, like vision or memory or conscious experience. A visual stimulus will have causal effects that percolate all through my brain and body, but only a subset of those effects are the visual processing itself. We thus need to carve out a subset of central brain events when we study emotion: they are those that explain how context-dependent stimuli lead to emotional behaviors under a functional description. We don't yet know exactly where in the brain this happens, but we know it's not everywhere, and we have a lot of data so far telling us where to look (the topic of the upcoming chapters).

As we discussed in the previous chapter, our view is that emotions evolved to deal with recurring environmental challenges that make particular computational demands, and that can be classified as requiring particular functional features in order to produce an effective, adaptive response. For instance, particular types of threats all require rapid deployment of a coordinated series of responses in order to avoid the danger, like the detailed scheme we showed in box 2.3 and the sample biology experiment we saw in figure 2.5 in the previous chapter. The particular responses required may well depend further on the type or proximity of the threat. There is an emerging picture of partly separable neural systems for responding to the threat associated with a predator, a conspecific, or another type of possible environmental danger (inclement weather, falling off a cliff, and the like [Gross and Canteras 2012]). So not all situations we would intuitively think of as "threats" may be treated as identical by the brain (see box 8.2 for some actual dissociations in rare human patients).

This may make it seem as though the central state of "fear" associated with processing threats would thus fractionate into a very large number of different adaptive problems. But in fact, these problems cluster around a much smaller number of categories or dimensions. For instance, threat due to a predator has a few, fairly well-defined variables that characterize the situation. There is the imminence of the threat (is the predator far away and has not detected you, or closing in, or about to pounce), and the availability to cope with the threat (are you stronger or weaker than the predator, is there a hiding place, are there others around who could help you) (see box 2.3). One way to think of this situation is thus as a decision-making problem in a relatively low-dimensional space defined by the values of these variables: this space captures the majority of situations in which you are facing a dangerous predator.

There will of course be further idiosyncratic details of any particular threat scenario for which emotions per se do not have a prepared (that is, innate) reaction, or not even a learned reaction. That is why emotion states must engage many other processes in order to further help the person or animal to deal with the threat. For instance, the precise path to take in order to flee from a predator may require substantial further processing in order to anticipate how I could outrun the predator, what the terrain is like, what obstacles are in the way, whether I should turn left or turn right, and so forth. The emotion state by itself is not responsible for the exact details of that behavior in all different circumstances. We stress this obvious point to acknowledge that emotions are only one of many states that contribute to an animal's behavior. Just like you cannot have behavior only by having attention, or only by having vision, or only by having memory, you cannot have behavior produced only by emotion. They all have to work together.

Exactly how emotion fits in to the economy of an entire organism's repertoire of different types of states and processes is a huge question. Answering it would amount to a full architecture of the mind, and for the neuroscientist, a story of how the brain generates the mind. Needless to say, that is not our task here. One can still make progress by bracketing these details. We can investigate emotion while still acknowledging that many nonemotional enabling (or "permissive") processes also need to occur. In experiments, we often do this by holding other processes relatively constant, or by measuring them as covariates. Just like it did not make sense to think of the amygdala in a dish as instantiating fear, in isolation from the rest of the brain, it also does not make sense to think of all visual processing and motor processing and attention and memory as instantiating fear. Emotions are processed by brain systems that are larger than a single structure (like the amygdala) but more circumscribed than all of the brain.

When we colloquially speak of somebody "consumed by anger," or think of an animal's behavior being "taken over" by fear, this does not mean that all of the behavior can now be explained just by fear. It simply means that many additional processes have been *recruited* in the service of the state of fear.

However, there do seem to be certain kinds of states that interact more intimately, or even overlap, with emotions. What about states such as motivation, arousal, and drive? It may turn out that these states should be thought of as dimensions of an emotion state. Indeed, many theories of emotion include one or more of these other types of states as properties of an emotion: it motivates behavior, and varies in arousal level, for instance. But it might also turn out that motivation and drive are independently useful ways to categorize other internal states of an organism, even though they overlap to some extent with emotion states. We see no real problem with this—one can choose to taxonomize the states of an organism in many different ways, and these need not be mutually exclusive if they are scientifically useful. I might give you an address in terms of a street number in a city, or I could give you GPS coordinates that your cell phone can home in on. These descriptions overlap, but neither one eliminates the other because both can be useful under different circumstances. If it turns out to be useful to speak of motivational states and drive states, under certain circumstances (see chapter 5)—so be it. This book is about emotion states, and we hope that by the time you have finished reading it, it may be more apparent how emotions differ from other states. To begin with, we will now list some properties of emotions that indeed set them apart from most other states.

Building Blocks versus Features

As we have already discussed, emotions should be thought of as adaptive functional states that are at a level of complexity intermediate between reflexes and volitional deliberation or calculation (although both reflex-like responses and volitional deliberation may be recruited into an emotional response). That is, emotions are, on the one hand, evolved packages of functional adaptations that are more encapsulated, and more domain specific (box 3.1), than volitional deliberation; yet, on the other hand, emotions are considerably more multifaceted and flexible than are reflexes. The emotion properties we will discuss reflect this level of behavioral control; for instance, emotions show persistence and learning as hallmarks of flexibility, yet show automaticity and prepotent control over behavior as hallmarks of their nondeliberative nature.

While there is no black-and-white dividing line, one can roughly partition the properties that emotion states exhibit into two classes: what we call building blocks and features. We think of building blocks as those properties of an emotion that are more essential, and more basic. All emotion states have most of the building blocks, and we can find precursors to emotion states in simpler organisms that already show many of the properties of building blocks. Features, on the other hand, are more elaborated, derived, and variable properties of emotions, and not all emotions have them. To use an automotive analogy, wheels are a building block, while air conditioning is a feature.

A good example of a building block would be valence. All emotion states are related to evaluating a stimulus as good or bad, as something to be approached or avoided. In psychological approaches that focus on the conscious experience of emotion, valence would correspond to how pleasant or unpleasant the emotion feels. Although we think that it is premature to commit ourselves to stating necessary properties of emotions at this stage, valence is a good candidate for a property without which you probably cannot have an emotion state.

A good example of a feature would be social communication. This is certainly a very prominent property of emotions, especially in mammals, but it is likely more recently evolved, and it may not be present in all cases. Even one and the same emotion state might play a functional role in social communication in some circumstances, but not in others. I could be in a state of fear if about to fall off a cliff when by myself, but I could also be signaling fear to another person if confronted with an aggressive burglar. Whereas valence is a building block that is always a property of an emotion state, social communication is a feature that is a more variable property only seen in some animals on some occasions.

BOX 3.1. Domain specificity.

Domain specificity refers to processing that operates on a restricted domain of stimuli, or even only on a very specific type of stimulus. For instance, face perception is thought to be relatively domain specific in that there are psychological and neurobiological mechanisms specialized to operate only on faces, and not on other visual stimuli. Similarly, language processing operates on specific kinds of auditory input. Where might emotions fall? Are particular emotion categories, like fear, disgust, or sadness, specialized for particular domains of stimuli in the same way that face processing and language appear to be?

Although the term *domain specificity* is used in a number of different ways, perhaps the clearest usage goes back to a technical definition of another term: *modularity*. In his book *The Modularity of Mind*, the philosopher Jerry Fodor first listed the features that he thought a modular system needed to have (Fodor 1983), not all of which are still agreed upon in modern discussions of the topic. However, there is one feature that seems to be the sine qua non of modularity, which is domain specificity. The idea was simply to describe the range of inputs that an information processing system could have access to.

In the case of the knee-jerk reflex, the range of inputs is extremely narrow: that reflex cannot decide to move your knee based on what you believe about the world, based on what you see, based on what you hear or taste or smell. It can't even incorporate information from touch elsewhere on the body. It only gets input from the stretch receptors in your knee. On the other hand, deliberating whether to go out for dinner tonight can take any input into consideration whatsoever—how hungry you are, what the menu of the restaurant looks like, what your wife says, and so on and so on with no obvious boundary. So the knee-jerk reflex is highly modular and domain specific, whereas deciding to go out for dinner is not.

Emotions, once again, are in the middle. An emotion state like fear can take certain kinds of inputs, but not any input. Bitter taste, for example, is not the right input for fear, but is instead the right input for disgust. A looming visual stimulus is the right input for fear, but not for disgust. There is much more flexibility than with reflexes, because there is learning—so you could try to train an animal, or a human, to be afraid of a bitter taste by pairing the taste with electric shock in a Pavlovian fear-conditioning paradigm. But not all stimuli are equally learnable in this way. Certain kinds of stimuli are much easier to learn to fear (and indeed many people fear them without having to learn to fear them at all): we can more easily be afraid of snakes and spiders than of flowers and paintings. So there seems to be at least a degree of domain specificity that would help to define emotion categories.

There is an alternative explanation to some apparent cases of domain specificity. For instance, the domain specificity of face processing is in fact debated (Kanwisher 2000; Tarr and Gauthier 2000). Face processing, like emotion processing, is sufficiently "central" that very many different kinds of inputs have access to such processing (unlike the case for reflexes). This makes it a challenge how to explain domain-specificity for these central cases, since it cannot arise simply from restricted sensory input, but must reside in some features of the central processing itself. For faces, it has been argued that fine-grained classification and expertise are information-processing features that generate domain specificity (because they apply mostly to how faces need to be processed). For emotions, a similar list of features may provide such an account (e.g., those we list in figure 3.2). Future work will need to consider a more flexible definition of "domain specificity" that considers how this may arise on the timescales of evolution, of development, or even of experience with domains of stimuli that make specific processing demands during one's lifetime (Spunt and Adolphs 2017b).

A Provisional List of Emotion Properties

Let us turn now to discussing the properties of emotions. What do we mean by this? One way of thinking about emotion properties is very practical: they are those processing features that you'd look for in the brain, if you are a neuroscientist, in order to discover putative emotion states. They need to exhibit certain properties, without which they could not qualify as implementing an emotion state, because they would not be able to carry out the functional role of that emotion.

Figure 3.2 provides a provisional list of the properties of emotions, some of which we discuss further in the remainder of this chapter. Several things should be noted about figure 3.2. First, it is certainly not intended to offer anything like a "theory of emotion" or a definition of emotion. It does not claim to be a complete list, and it may even be that some of these properties should be omitted eventually. Instead, it is intended to illustrate the approach to investigating emotion states in general that we believe will be most useful: begin by listing operating characteristics of emotion states in virtue of which they are able to carry out their functional roles. The list in figure 3.2 is generic for all emotions. If one is investigating a specific emotion, like fear, or disgust, or shame, specifically, one might begin with a more detailed list of the functions of each of these emotions (for instance, perhaps something like what we showed in figure 2.4 in the case of fear). We are shying away from undertaking this latter task in this book since, as we mentioned, we feel it may be premature to focus on specific emotion categories (except by way of example).

Scalability. An emotion state can scale in intensity. Importantly, parametric scaling can result in discontinuous behaviors, such as the transition from hiding to fleeing during the approach of a predator (cf. Box 2.3). Intensity is often conceptualized as arousal, although these two are not the same thing.

Valence. Valence is thought by many psychological theories to be a necessary feature of emotion experience (or "affect"). It corresponds to the psychological dimension of pleasantness/unpleasantness, or the stimulus-response dimension of appetitive vs. aversive. (But, again, these two are not the same thing.)

Persistence. An emotion state outlasts its eliciting stimulus, unlike reflexes, and so can integrate information over time, and can influence cognition and behavior for some time. Different emotions have different persistence. Emotions typically persist for seconds to minutes.

Generalization. Emotions can generalize over stimuli and behavior, much of which depends on learning. This creates something like a "fan-in"/"fan-out" architecture: many different stimuli link to one emotion state, which in turn causes many different behaviors, depending on context. Persistence and generalization underlie the flexibility of emotion states.

Global coordination. Related to the property of generalization is the broader feature that emotion states orchestrate a very dense causal web of effects in the body and the brain: they engage the whole organism. In this respect, they are once again differentiated from reflexes.

Automaticity. Emotions have greater priority over behavioral control than does volitional deliberation, and it requires effort to regulate them (a property that appears disproportionate, or even unique, in humans.)

Social communication. In good part as a consequence of their priority over behavioral control, emotion states are pre-adapted to serve as social communicative signals. They can function as honest signals that predict another animal's behavior, a property taken advantage of not only by conspecifics, but also predators and prey.

FIGURE 3.2. The properties of emotion states. This important table lists properties of emotions that we will continue to discuss in this book. Some of the properties listed near the top could be thought of more as building blocks of an emotion state; others listed near the bottom could be thought of more as features (see text for details).

Similarity Structure

The properties shown in figure 3.2 can be used to characterize a particular emotion state and can serve to distinguish it from other emotion states, as well as from reflexive behavior. Some instances of an emotion have more or less positive valence than others; some persist longer than others; some play more of a role in social communication than others. If we consider all these properties as dimensions on which an emotion state can vary, they specify a dimensional space in which we can map emotions. A simplified example of how this might work is shown in figure 3.3.

The concept of a dimensional space within which to characterize emotion states introduces us to one of the most fundamental properties of emotion states: different emotion states can be related to one another as more or less similar. In a way, this is a truism. Absolutely everything can be related to everything else this way. As we will see, this is a major approach used also in studies that employ functional neuroimaging to investigate human emotions (chapter 8). One key question is, what metric, what dimensions, are being used to make the similarity comparison?

If I ask you which of two pairs of numbers have greater similarity, 2 and 3, or 2 and 326, what would you say? Almost certainly, you'd say 2 and 3 are more similar, and your intuitive metric would simply be their numerical difference. You can graphically plot them in a one-dimensional space, a number line, and you can see that 2 and 3 are closer together than are 2 and 326. So that one is easy.

Now let's consider a harder case. Which of these animals seem more similar to you: cats and dogs, or cats and eagles? While a modern biologist would answer this question using comparative DNA sequencing, since most people don't have access to a DNA sequencer, they would use their general knowledge of biology, or their intuition. You would probably answer quickly that cats and dogs are more similar, but if you used your intuition it might be harder to specify how you did it. The reason is that the similarity space in which you are comparing these animals is generally more high-dimensional. It's not a single number line: size or weight seem insufficient as metrics to judge similarity. You

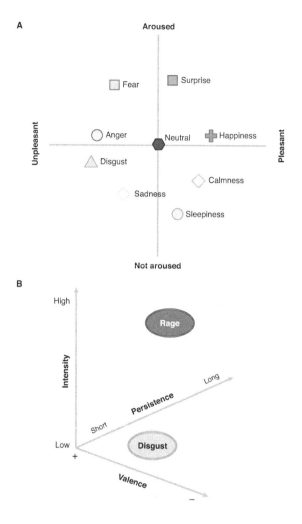

FIGURE 3.3. A dimensional space for emotions. (A) The most popular dimensional space used in psychology, which plots emotions in a two-dimensional space of valence and arousal. It often produces a circular arrangement from which the similarity structure of emotions can be easily seen, as noted by the psychologist James Russell (Russell 1980). (B) This figure is intended to show conceptually how emotions might be characterized without needing to classify them as "anger," "fear," disgust," or any other category for which we happen to have words in English. Instead, they would just be points in a multidimensional space (we only show three dimensions here for ease of visualization). For example, it might turn out that all the instances of emotions related to attacking and defending are clustered in a certain region of that space, such as the red region denoted by "rage." Then again, it might well turn out that we find clusters that do not correspond so easily to terms we have in our current vocabulary for emotions. It is even possible that they do not cluster at all. A dimensional space can accommodate all of these possibilities; it is entirely feasible to investigate emotion without using verbal labels to describe *which* emotion you are investigating. Reproduced with permission from Anderson and Adolphs 2014.

have to think about other attributes, like the presence of fur, the likelihood of being a pet, having four legs, walking on the ground, etc., etc. So cats and dogs and eagles can be compared in a higher-dimensional similarity space, even though we are not sure what the dimensions of that space are. Yet, we can compare them, and once again the graphical representation would show a geometry like what we saw with the numbers: cats and dogs are closer to one another in this high-dimensional space than are cats and eagles.

Arousal and Valence

Emotions can also be compared to one another in terms of similarity, but, as with the cats, dogs, and eagles, it is not clear which dimensions to use in order to make the comparison. Most commonly, emotion researchers have used two dimensions since these seem to capture our intuition about similarity relationships between different emotions remarkably well. Those two dimensions are something like arousal or intensity (what we called "scalability" in figure 3.2), and valence (or pleasantness). That is, all emotions are graded or scalable, and all emotions are more or less pleasant or unpleasant. The two dimensions of valence and arousal capture people's ratings of emotions remarkably well, so at the psychological level, and in particular at the level of how people conceptualize emotion experience (feelings), there are good reasons for quantifying the similarity between emotions in this way. It does not matter too much if you show participants facial expressions of emotion, or you just give them words or stories that evoke emotion concepts—they can be ordered in terms of their similarity in this two-dimensional space. Psychologists refer to the two-dimensional space that characterizes aspects of our experience of emotion (or our concepts thereof) as "affect," and in most theories, it forms a sort of core on which all emotional feelings can be elaborated (Russell 2003).

However, this 2-D space of valence-arousal is derived primarily from human subjects, and primarily from judgments that they give *about* emotions. It is not derived from behavioral or neural measures of actual emotion states, which are the ones that the emotion scientist may want to use. There are alternative ways of mapping emotions in a 2-D similarity space that uses different kinds of data. For example, you could also base the scheme on stimulus-response relationships, rather than on properties of their concepts or their conscious experiences. Such a scheme was in fact proposed by the neuroscientist Edmund Rolls, in his book *The Brain and Emotion* (Rolls 1999). Rolls argued that emotions can be conceived of as states elicited by the administration or withholding of reward or punishment. This also generates a 2-D similarity space, but the axes now correspond to the administration (or withholding) of reward (or punishment). While quite different from the standard psychological affect space that has arousal and valence as its dimensions, this is another example of how an emotion state (or, in this instance, the stimulus-behavior mapping that emotion states accomplish) can be represented in a relatively low-dimensional space. A given specific emotion can be situated in either of these two-dimensional spaces. Fear, for example, would be high-arousal, negative valence. It would also be a state caused by administration of punishment (or by the anticipation of punishment).

None of these two-dimensional schemes capture all the variance among emotions, and it seems certain to us that schemes with a larger number of dimensions will be required. We do not know yet how many dimensions there should be. One guide may be to make the dimensions equivalent to a subset of the particular properties we list in figure 3.2, so that emotions are described not only by valence and arousal but also by how long they persist in time (something like a decay constant), or how much they can generalize across stimuli (which may amount to something like the extent to which they are based on innate triggers versus learned ones). It would require some ingenuity to come up with multiple dimensions that each permit a scalar quantity to be mapped; ultimately, we need to be able to put a number on the value that any emotion takes on any given dimension, and such quantification may be difficult at present for many of the properties we list. We will return to similarity spaces in chapter 8, when we discuss the neural basis of emotion from fMRI data, and we will return to valence in chapters 9 and 10, since it is a feature that is particularly prominent in the conscious experience of emotion.

Persistence

We just discussed scalability and valence when we sketched how emotions can be characterized and compared in a similarity space. But scalability and valence are independently important properties of emotions that we can measure, and that enable particular functions. Scalability in particular distinguishes emotion states from simple stimulus-response reflexes, which tend to be all or none rather than graded. We now turn to brief discussions of the remaining properties listed in figure 3.2.

An important aspect of flexibility is enabled by the ability to integrate information over time and to combine multiple sources of information. This integration function requires the property of *persistence*. If an emotion state persists for some time, it can accumulate sensory information over time and from multiple sources; it can interact richly with other internal states; and it can have a prolonged effect on behavior. This is typically what we observe: if somebody is in a state of fear, they are in that state for some time, and the way in which stimuli influence their behavior is colored by the persistence of that fear state. We see the same regularly in animals: a mouse in a state of fear that has just escaped from a predator may hide in a burrow for quite some time before emerging, cautiously and tentatively. The state of fear has a long-lasting effect on the behavior. Moreover, different emotions seem to have different degrees of persistence: surprise and fear are relatively fleeting, whereas sadness tends to persist for longer. Persistence may thus be one useful dimension that maps different categories of emotion. The property of persistence has been used to help characterize emotion-like states even in flies (Gibson et al. 2015). Emotion-like states that persist for hours or days or longer are usually classified as moods (but there are other distinctions between moods and emotions also; see box 3.2).

BOX 3.2. Moods.

While emotion states do have the property of persistence, they generally do not persist for long once the situation that triggered them has been resolved. Once you've escaped the bear, your fear gradually declines (and might be replaced with relief instead). But there are also emotion-like states that last much longer.

Moods are like emotions in many respects, and are typically accompanied by feelings that are similar. However, they last much longer than emotions—anywhere from hours to years. Moods seem more prominent in humans than in animals, perhaps reflecting our propensity to think about memories and plan for the future more so than animals do. Whereas emotions function to cope with acute, current situations, moods often function to cope with events in the past or in the future—and indeed can be induced by representations of events that are entirely imagined.

Although moods and emotions share some aspects, they are sufficiently distinct that moods should not simply be considered as temporally extended emotions. According to some emotion theories, we are constantly in an emotion state, which just waxes and wanes as we encounter stimuli. In our view, emotions only occur when specific functional challenges are encountered, and you are not generally in an emotion state if you're just sitting at home having breakfast (but you could be in a mood).

Unlike emotions, moods often have no clear trigger (perhaps one reason for their long persistence). They also seem to involve effects on cognitive processes more than effects on behavior. We generally do not start running away or performing particular goal-directed behaviors as the result of a mood. But we do start thinking about different subject matter, and paying attention differently. These effects of moods on other cognitive processes like memory, attention, and decision-making are a large topic of study in psychology.

Of the emotion properties we listed in figure 3.2, moods share persistence, scalability, valence, generalization, and automaticity. They don't seem to function much in social communication.

One could extend the temporal scale from emotions to moods all the way to personality traits. Emotion-like personality traits have also been a big topic of study in psychology. People, and to some extent animals, show dispositional affect, which predisposes them to exhibit certain emotion responses. Some people are generally made more happy or more sad or more afraid by otherwise equivalent stimuli. These are aspects of individual differences that are more stable across time, that probably have a substantial genetic component, and that are important topics of study in psychopathology (see box 2.2).

State persistence is a very useful property from the formal perspective of decision-making. One of the most influential classes of computational models, which can describe how the accumulation of sensory evidence results in a choice, are so-called drift-diffusion models, originally conceived by the psychologist Roger Ratcliff (Ratcliff and McKoon 2008). There are many formulations of such accumulator models, which are often based on reaction-time data in psychology, but figure 3.4 shows the essence of a drift-diffusion model.

In figure 3.4, we can see that a response may be evoked only after some time. You can easily try this yourself: sit out in the dark in an unsafe alley and listen to the little rustling sounds and footsteps around you. Even if nothing changes in the environment, it will only be a matter of time before you become so afraid that you need to hide or run away. Your brain is accumulating sensory information that informs it of potential threats. If the sensory evidence is large, such as seeing a tiger right in front of you, your decision is rapid, since the decision threshold is crossed almost immediately. If the sensory evidence is low, such as your dark alley with little rustling noises, your decision is slow, since it takes a while for sufficient sensory evidence to have accumulated to cross the decision threshold. If your decision threshold is low (the horizontal black line that is crossed by the blue curves), you will make a behavioral decision sooner, and this might correspond to an individual who already has a trait of being anxiety prone. If the threshold is really low, you will make a decision very quickly, but will also have more false positives—you might run away from almost anything, so easily are you scared (see box 2.2).

This type of model has been quite powerful in explaining how the brain makes behavioral decisions, because it has several free parameters (which can be fit from data) that describe the mechanism of the choice

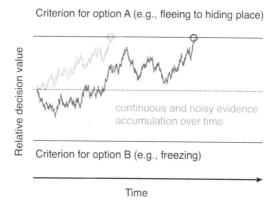

Criterion for option A (e.g., fleeing to hiding place)

Relative decision value

continuous and noisy evidence
accumulation over time

Criterion for option B (e.g., freezing)

Time

FIGURE 3.4. Drift-diffusion model. In this schematic, a behavioral (or neural) decision is modeled as the noisy accumulation of a signal, or a set of signals, over time. In the figure, an example is depicted in which one must make a behavioral decision based on sensory input, such as detecting a predator. Over time, sensory signals may accumulate to cross a decision threshold bound and result in a response. The light blue line might correspond to the decision to flee if you are in a dark alley and hear footsteps approaching you. The dark blue line might correspond to the more delayed decision to flee if you are just in a dark alley getting more and more scared, but not hearing any footsteps approach.

process. How high the threshold is, what the starting point is, the rate of accumulation of evidence and whether the decision bounds are fixed (as in figure 3.4) or collapsing (as when you will make a decision after some fixed time, regardless of the amount of sensory evidence), are all aspects of the model that can be considered, and that will describe aspects of the behavior (how quickly you run away, how many false positives you make, how sensitive you become to additional information over time, and so forth). Such models have also been used to probe brain responses, and they offer a powerful framework for incorporating the persistence feature of emotions.

Persistence can also be thought of simply as a scalar quantity corresponding to something like a decay constant. Just like any radioactive element can be identified from a unique fingerprint, its rate of radioactive decay, every emotion state might have a decay constant. Perhaps we can distinguish different categories of emotions, in part, by their different temporal decay constants. In radioactive decay this is easy, since the decay is described by an exponential function; no matter at what point in time or for how much material you do the test, comparing across time points will allow you to fit a unique decay constant that is a signature for your element. It would be a wonderful tool if we found something similar for emotions!

You might think that the persistence of emotion would be complicated by, or indeed confounded by, other aspects of memory. Surely, how afraid I am will depend on whether I remember being afraid, or remember what triggered my fear in the first place. Indeed, these are complexities that would need to be taken into account—but they also can be stripped away to demonstrate that emotion states have persistence in humans independently of whether we consciously remember anything. Such an experiment was done by the cognitive neuroscientists Justin Feinstein, Melissa Duff, and Daniel Tranel, who tested patients who were amnesic (Feinstein, Duff, and Tranel 2010). In that study, the patients had damage to the hippocampus, a brain structure necessary for encoding new memories. The patients could not consciously remember any event for longer than about a minute. The researchers showed them movies that evoked strong emotions, such as anger, fear, happiness, and sadness, and the patients all felt strong emotions as they watched the movies, just like healthy individuals. Several minutes after they had watched these emotion-inducing movies, the patients were asked how they felt, and what they remembered. It turned out that the amnesic patients still felt strong emotions, especially sadness, for quite some time after having seen the movie (as would everyone)—but in their case, they could no longer remember why. They did not remember having seen the movie and could not provide any explanation for why they felt sad. This study demonstrates that the persistence of emotion states is not simply a consequence of our ability to consciously remember what emotion we felt, or to remember the events that elicited the emotion. Emotion states have an intrinsic persistence, and this persistence has a functional role in their ability to integrate sensory inputs for some duration, and to influence cognition and behavior (and other emotions) over some period of time.

Generalization

A very important aspect of emotion, and another one that distinguishes emotions from reflexes, is their ability to generalize over stimuli. This generalizability is one of the most poorly understood properties, since it goes well beyond categorization or abstraction as typically studied in perceptual systems. Given the highly varied and multimodal sensory inputs that can carry information relevant to a particular emotion (for example, predictors of a threat that could induce a state of fear), there is no simple formula that determines which stimuli cause an emotion, let alone which stimuli cause one type of emotion rather than another.

The most common way that scientists have dealt with this complexity is to push it into the domain of "context effects." That is, it has often been assumed that there is still a relatively small set of stimuli whose sensory features cause emotions, but that this can be modulated or gated by the context in which the stimuli occur. For instance, your fear of a snake would depend on whether you are alone in the bush, whether you are with other people who may protect you, or whether you see it in a pet store behind glass or just happen to know that it is defanged. As is apparent from these examples, so-called context effects quickly become so complicated that they can incorporate most of cognition. There is now a large subfield in affective neuroscience devoted to emotion-cognition interactions that study exactly this topic (Pessoa 2013).

Another way to confront the complexity due to generalization is to situate the stimuli within a richer model, ideally a model that can be formalized functionally and computationally. For instance, the context in which you see the snake (bush, with others, in pet store, and so on) could be mapped onto a scheme such as the one we showed in figure 2.4.

Generalization is closely linked to learning, which is a huge topic in emotion research we mention again further on. Most of the stimuli in our lives that cause emotions have acquired their causal efficacy through experience. The best understood example of such associative emotional learning is Pavlovian fear conditioning, which we will discuss in considerably more detail in chapter 6.

There is also generalization in the behavioral reactions caused by an emotion. While Pavlovian unconditioned responses such as freezing or salivating (see chapter 6) may constitute a relatively small set of behaviors, instrumental behaviors that are learned constitute an open and large set of behaviors that can be flexibly linked to an emotion.

There are important species differences in generalization due to learning. While all animals show instances of associative emotional learning similar to the previous examples, only certain species appear to show learning that takes place through observation rather than through direct experience. You generally do not need to actually experience dangerous or disgusting things yourself in order for the stimuli to cause those emotions; you can learn by watching other people (and so can many animals). For instance, neither one of us has ever been threatened with a gun, but both of us are afraid of guns—because we have seen so many examples of other people threatened by guns (primarily in news media). There may be one type of learning about stimuli unique to humans: simply being told about them. The extensive social learning due to observation and communication through language likely makes human emotions considerably more generalizable than the emotions of other animals.

Together with persistence, generalization offers enormous flexibility in how emotion states link stimuli to behaviors. The causal architecture links very many stimuli, most of them through learning, to very many different behaviors (a "fan in, fan out" architecture). For example, the same emotion state (fear) can be expressed by a variety of different movements and postures (freezing, flight, vocalizations, facial expressions), depending on the proximity of the threat and the context and the animal's past experience. This achieves functional cohesion across a large diversity of situations; emotion states are portable mechanisms for achieving the same function (for example, coping with a threat, avoiding a poison) across an unbounded number of different circumstances with which the world may challenge us.

Global Coordination

Another property listed in figure 3.2 is global coordination. Emotion states are distinguished from most other central states by the extent to which they causally interact with other internal states, including states of the body. Emotions not only influence behavior and cognition, but also prominently influence endocrine and autonomic responses such as changing our blood pressure, influencing our digestion, and altering hormone levels. The global engagement of many organism variables is probably an important clue to the kinds of environmental challenges that emotions evolved to deal with: they require a coordinated response involving the whole organism.

Persistence, generalizability across multiple behavioral outputs, and global engagement of the organism all require another key property of emotions that is often stressed in emotion theories: coordination. The causal consequences of emotion states need to be cohesive in some way, and so they must coordinate different aspects of cognition and behavior in time. When we begin discussing neural mechanisms in the subsequent chapters, this theme will become particularly important, since it will become even clearer that the causal consequences of a central emotion state are very diverse. There are effects on our body (changes in heart rate, breathing, blood pressure, pupil dilation, salivation, hormones—all mediated through the autonomic nervous system) as well as changes in behavior (running, lunging, vomiting, and such—all mediated through the somatic nervous system). In addition to these bodily effects, there are of course effects on many cognitive processes as well, as we have repeatedly stressed.

For instance, a state of fear typically includes EEG arousal (generalized arousal that can be measured from the brain using electroencephalography), pupil dilation, increased gamma motor neuron activity (muscle tone), and increased respiration, among many other effects. How are these all caused and coordinated? Clearly, emotions require some central coordination and control in order to produce such concerted behavioral responses. Some of this is mediated by known anatomical projections—for instance, the central nucleus of the amygdala projects to brainstem and hypothalamic target regions that in turn mediate each of these behavioral components through a complex circuitry that is still being worked out in detail (figure 2.1).

The picture suggested in figure 2.1, while true to some extent, is likely too simple to explain the coordination that we see, since an emotion state will be much more distributed than emanating from a single structure like the amygdala. Also, different subsets of behavior can be seen on different occasions, or emerge at different points in time. So, clearly, the full story about how multicomponent coordination is achieved is going to be complicated and is going to differ to some extent for different emotions and different species. Nonetheless, it is one global feature that we can look for in the brain; like persistence, multicomponent coordination is one functional feature that we can use to search for a neural mechanism of an emotion state in the brain; perhaps it involves synchronized oscillations across different brain regions, or other long-range mechanisms to link activity in different areas. It is also, once again, a feature that distinguishes emotions from reflexes; the latter do not need to coordinate a large number of multiple components, because they don't have a large number of components. If your knee jerks, you don't need to coordinate your blood pressure, heart rate, hormone levels, or pupil dilation along with it. The multicomponent nature of emotions derives from the fact that they are global organismic states (LeDoux 2012); their function relates to an entire organism in a complex environment, not merely the movement of your leg as in the knee-jerk reflex (although, of course, moving your leg could be a component of an emotional response, and the knee-jerk reflex would help support your running away from the bear).

There are multiple system architectures that could achieve such global coordination. At one extreme, every behavioral component could be entirely separate, and there could be a completely distributed system of communication between all the separate components; for some, one component facilitates another, for others, one component inhibits another. With the right connectivity, such a distributed network could still achieve a coordinated response.

At another extreme, there could be a single command neuron that orchestrates all the behavioral components, in a strict hierarchy. Examples of both types of mechanism have been found, and it seems clear that the answer is generally somewhere in the middle. Much the same is true of structures like the internet: there is not a single server that controls everything, nor is control completely distributed. Instead, there are routers and hubs that organize the internet at an intermediate level of control between a strict hierarchical control and a fully distributed control.

It is important to emphasize again that the causal control exerted by an emotion state is distributed in time as well as in space (figure 3.1).

This may seem obvious, since emotional behaviors are themselves distributed in time. We need to be able to explain the temporal structure of these behaviors: some classes of responses are reflex-like in their rapid speed and shallow depth of processing (for example, immediately shrinking back from an attacking bear), whereas others are deliberation-like in involving lots of slow integration and predictive processing (for example, planning how to avoid or escape from a bear that is some distance away in a complex environment with several possible escape alternatives). Although we focus on those aspects of an emotion state that lie intermediate to these two extremes (for example, the bear is coming toward you and you need to escape now), emotion states often involve a large temporal range of sensorimotor processing, from very fast to very slow.

How the distributed neural components that implement an emotion achieve cohesive control over all the causal effects of an emotion is still quite a mystery. Indeed, it is a mystery that we will need to understand if we want to design emotional robots. It involves a difficult optimization decomposition problem: each component is implementing an aspect of the emotion, but many components (distributed in both space and time) have causal effects on a shared set of effectors. The distribution of control, together with the fact that the multiple components necessarily interact, is what makes the analysis and architecture of the entire system so complex to understand.

Just like we had drift-diffusion models as one possible mathematical tool that we might use in understanding and modeling the persistence feature of emotions, there are also powerful mathematical models that can be applied to neurobiological data to understand the architecture that explains how emotion states cause coordinated behavioral components. One type of model is called a directed graph, which is simply a mathematical formalization of causal influence: one brain region (or neuron) might project to another brain region (or neuron) and cause a response therein (figure 3.5). You could think of it as the wiring diagram in the house where you are trying to figure out how the light switches work. The two regions or neurons would correspond to two nodes in the graph, and the causal influence of region A on region B would be depicted as a directed edge (an arrow in the figure).

Many experimental neurobiologists make such models informally, and cognitive psychologists do too, but they are usually extremely simple. The field of causal discovery can use experimental data to generate graphs with thousands, or even millions, of nodes (Eberhardt 2017). There are even techniques for recommending what experiment you should do next (which node you should experimentally manipulate) in order to help determine the unique architecture of the graph, at least in mathematical theory. With more and more large datasets available, and more and more powerful computational methods for finding the right graph, this is likely one field that will contribute substantially to our understanding of the causal structure of an emotion state and how it achieves its multicomponent control and coordination. The main limitations to this kind of analysis are having enough data, and having specific enough tools for focal perturbation.

Automaticity

We will close this chapter with a final set of features. These are once again related to how emotion states function at a level of control and complexity that is intermediate between reflexes and deliberated actions, a distinction related to the concepts of automaticity and control.

The term *automaticity* has a historical formal usage from psychology that has had a very long success. Still to this day, human cognition is generally thought to exhibit features of two different systems (although there is debate about whether they actually comprise systems of any sort). These go by various names and attributes. The Nobel laureate Daniel Kahneman referred to them as "thinking fast" and "thinking slow" in his eponymously titled book (Kahneman 2011). He also refers to them, as many economists do, as "System 1" and "System 2," which are merely labels that subsume the many features these two cognitive systems are thought to have. Figure 3.6 lists some of these features. Some researchers debate whether all these features actually apply, and whether they are tied together in any interesting way—it may not be the case that there is a "System" corresponding to the two lists shown in the figure. Nonetheless, several of the entries under System 1 often co-occur in studies: notably, emotion and automaticity.

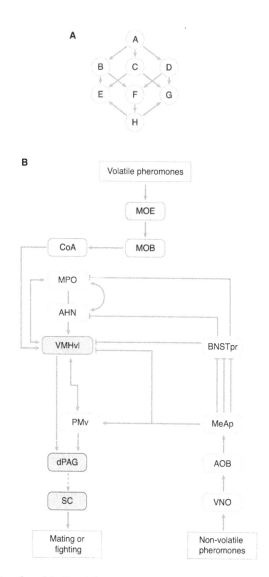

FIGURE 3.5. Causal models. *Top*: Schematic of a very simple type of graph, a directed graph that might be used to model how a stimulus, A, causes a behavior, H, via a number of intermediate stages in the brain (B–G). *Bottom*: More typical example from neuroscience. The lines denote effects among various brain regions in the mouse that mediate how the processing of odorants causes particular behaviors. This schematic is based on actual data and amounts to a hypothesis about the type of causal influence (for instance, excitatory or inhibitory). It is important to note that causal models like these are useful at all levels—from the broad conceptual level often seen in psychological box-and-arrows models, to the macroscopic neural systems level typically investigated with human lesions or fMRI, to the circuit and cellular level that can be dissected in some animal models. They vary from highly speculative hypotheses, to detailed causal connections supported by direct anatomical connections. Modified with permission from Anderson 2016.

You will see in this list the original entry from Shiffrin and Schneider (1977), who were the first to propose the dual-process scheme; this is the entry "controlled," which is contrasted with "automatic." By this, psychologists meant, roughly, those aspects of processing over which you have volitional, conscious control, versus those over which you do not. You will see several other attributes of System 1 that makes it look somewhat similar to the attributes of emotions—indeed, System 1 is sometimes associated with, or even identified with, emotions—but this list is not based on particularly strong data and should not be taken too literally, in our view.

One way of describing the way in which emotion states exhibit "automaticity" is with respect to their priority over behavioral control, a property that they share (but to a lesser degree) with reflexes. In general, if an emotion state can elicit a behavioral response, this will often happen without any further effort on your part. Indeed, you typically have to exert effort for the emotion *not* to cause the behavioral response. It takes an additional level of control *not* to punch the person with whom you are angry, and *not* to run away from a scary animal, and *not* to gag while putting a spider in your mouth. So it certainly seems as though there can be competition between systems vying for behavioral control in these examples.

However, the apparent competition between prepotent emotion and effortful volitional control does not necessarily need to arise from competition between two separate systems in the brain, as figure 3.6 would suggest. Returning to our example of drift-diffusion models (figure 3.4), it is possible to capture the apparent competition as unfolding within a single process—one that accumulates signals from two sources. One source may correspond to an "emotion-driven" signal, whereas the second may correspond to a "volitional control" signal. The mathematical formalism of a drift-diffusion model can accommodate both as signals that steer the decision-making process, only in opposite directions. Such an approach has in fact been applied to modeling human decisions that are driven by both emotional and controlled processes, such as altruistic decisions, where the weight that is put on how we feel about another person exerts an influence on the eventual decision to behave altruistically or selfishly (Hutcherson, Bushong, and Rangel 2015).

System 1	System 2
Automatic	Controlled
Heuristic	Systematic
Fast	Slow
Effortless	Effortful
Non-conscious	Conscious
Emotional	Rational
Implicit	Explicit
Reflexive	Reflective
Intuitive	Analytic
Parallel	Serial

FIGURE 3.6. Dual-process theories of cognition. There are many such schemes, but they tend to look similar, and tend to align emotion with "System 1." The basic idea is that the two systems vie for control of thought and behavior, and each can come into play depending on the circumstances. Although this is an appealing idea, the evidence that all of these properties actually go together, or belong to a "system" of some kind is not well supported. Nonetheless, emotion and automaticity tend to co-occur across many studies. It may be that many of the entries under System 1 evolved before those under System 2 (Bach and Dayan 2017). If so, and if System 1 corresponds to or includes emotion systems, it would support the idea that emotion states are evolutionarily ancient and widespread throughout animal phylogeny.

There is a close connection between decision-making and several of the properties of emotions (notably, persistence, global coordination, and automaticity; see box 3.3).

The automaticity property of emotions is related to older concepts in cognitive science that thought of emotions functioning as "interrupt" mechanisms when responses to a particular situation need to be prioritized. For instance, the Nobel laureate Herbert Simon wrote,

> If the organism is to survive, certain goals must be completed by certain specified times . . . the provision must be made for an interrupt system . . . all the evidence points to a close connection between the operation of the interrupt system and much of what is usually called emotional behavior. Further, the interrupting stimulus has a whole range of effects, including (a) interruption of ongoing goals in the

CNS and substitution of new goals, producing, inter alia, emotional behavior, (b) arousal of the autonomic nervous system, (c) production of feelings of emotion. (Simon 1967)

It is important to note that many examples of control over emotions typically pertain to adult humans. They are less pronounced for children, and may not have clear counterparts in animals, where the emotion often seems to fully control the behavior. From this perspective, the argument that emotions are unique to humans seems precisely backward; if anything, animals are more "emotional" beings than people. Conscious deliberation and emotion regulation, the ability to control the experience and/or expression of emotions volitionally, may be relatively unique to adult humans. Emotion regulation is a large topic in the psychology and neuroscience of emotion, in part because difficulties in emotion regulation are an important aspect of psychopathology (Ochsner and Gross 2005).

A very interesting line of research could examine possible animal analogs of emotion regulation—such as may be the case, for example, in a dog that is trained not to run away from a stimulus that it fears. This is much, much harder to achieve in nonhuman animals, and the kind of control over emotions in these cases may be fundamentally different from the more metacognitive volitional control in humans. Box 3.4 explains one scheme for thinking about human emotion regulation, from the work of the psychologist James Gross.

Automaticity is also more dominant as we go backward in human development. Adults have substantial control, children and infants do not. This finding has informed a large body of neuroscience research as well, since we know which parts of the brain change the most during development. A region called the prefrontal cortex, in the very front of the brain, develops latest—and is one of the main regions that implements control over emotions.

BOX 3.3. Emotion and decision-making.

Both behavioral and neurobiological studies of decision-making have recently been informed by mathematical models that describe the way in which organisms respond to stimuli so as to achieve adaptive outcomes. This field, which encompasses neuro-economics, decision neuroscience, and cognitive science more broadly, provides arguments strikingly similar to those we have presented so far in this book.

The basic argument runs as follows: first, actions that are optimal from a mathematical perspective are generally impossible for organisms to achieve—the computations required become far too difficult for any realistic situation. Second, there are, however, approximate heuristics that can be applied to particular, recurring patterns of environmental challenges. Third, these packages of adaptive heuristics, in many ways analogous to emotion states, solve decision-making problems using several mechanisms, ranging from simple Pavlovian responses to habit-based and goal-directed forms of instrumental control. From the perspective of this computational decision-making framework, emotions have been dubbed "control algorithms for survival" (Bach and Dayan 2017). Notably, this framework proposes that there are specific emotion categories, suited to deal with specific environmental challenges (e.g., fear, disgust, anger, and so forth—related to the concept of domain specificity that we discussed earlier), and also that there are continuous parameters in these states (analogous to many of the properties of emotions we list in figure 3.2).

Effects of emotion states on decision-making have been studied in a range of contexts. Most experiments induce states that are more akin to moods—relatively long-lasting, valenced states that can then bias decision-making over some time. For instance, humans show particular biases in the kinds of risks that they are willing to take in their choices, and these are influenced by their mood. Some people are more pessimistic, some people more optimistic, and the biased expectations of the outcomes of the choices they have guide the decisions they make (Sharot 2011).

Emotions bias decision-making in animals as well. Much like humans, many mammals show prolonged shifts in the kinds of risky choices they make, depending on whether they have encountered stimuli that triggered negatively or positively valenced states. Even bees have been reported to have something like a pessimistic bias on their decisions (see chapter 7). In that study (Bateson et al. 2011), the researchers shook the bees to induce a negatively valenced, anxiety-like emotion state. They then tested how the bees would approach ambiguous choices and found that the agitated bees behaved as though they expected a higher likelihood of a negative outcome—akin to "pessimism" in humans.

BOX 3.4. Emotion regulation.

Emotion regulation refers to the ability to have some degree of volitional control over your emotion state, the conscious experience of that state, and the behavioral and autonomic expression of that state. As already suggested by this broad range of phenomena where emotion regulation could come into play, regulation is often broken down in relation to the temporal evolution of an emotion. There could be regulation at the very point of induction of an emotion, for instance, by choosing circumstances and environments that will modulate whether and how a certain emotion is induced. For instance, if I know I am afraid of public speaking, I might avoid situations where I have to give public talks. That would be a very indirect, strategic way of regulating my emotion.

More internally, faced with a situation or stimulus that could induce an emotion, we are often capable of reinterpreting the situation so that its emotional meaning is changed. We often do this in coming up with alternative explanations for negative social emotions, for instance—we might feel sad or guilty at having failed an exam, but might convince ourselves that we really did our best and the exam was just too hard. This level of emotion regulation is called "reappraisal," since it is a reevaluation of a typically complex situation in light of its meaning for us and our personal goals—the picture of emotion induction that psychological appraisal theories have long championed. We will say a little more about appraisal theory in chapter 10.

Finally, there is the attempt to control the experience or expression of the emotion state itself—to literally try to feel less sad or angry or afraid. This is the hardest stage at which to try to control emotions, since they are already unfolding. Often, if you try really hard not to feel angry when in fact you already are angry, you only get angrier. It is also very exhausting to try to suppress emotional

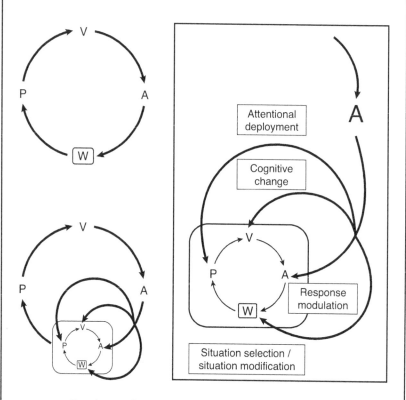

FIGURE 3.7. Emotion regulation as a metacognitive ability. Emotions can be broadly thought of as illustrated in the *upper left*: they involve a cycle between perceiving a situation in the world (P), assigning a value to it (V), and then motivating an action (A) that acts back on the world (W) to optimize the way an animal or person interacts with its environment. However, if we take a broader perspective, the world can include ourselves and our emotional reactions within it. From that broader perspective (*bottom left*) there is another, bigger, cycle that can act on the smaller one: we can perceive that our anger toward our supervisor is not working out so well, we can assign this insight a value on which to act, and the metacognitive action can then regulate our anger. This regulation, in turn can focus on multiple processes within the cycle (*right box*): it can involve redirecting our attention to change perception, it can involve reappraisal to change our valuation of what we perceive, it can suppress our emotional behaviors, and it can strategically select situations that we put ourselves in to help with our emotions. Modified with permission from Gross 2015.

experiences and behaviors at this stage of regulation. The ability to regulate emotions, at all these different levels, is thought to prominently involve the prefrontal cortex, a region of the brain that only matures relatively late in adolescence in humans.

One puzzle about emotion regulation is why it should be required in the first place. Aren't emotions supposed to be adaptive? Didn't they evolve to deal with problems we face in the environment? Yes, they did—but remember that we situated emotions in between the rigid reflexes and the full flexibility of volitional action. To some extent, emotions are domain-specific and thus unable to consider certain types of information that may be relevant when considered in a broader context (box 3.1). Emotions may respond to a stereotyped situation that confronts us, but broader considerations and thoughts about future consequences may suggest a different course of action would be better. Such broader considerations require some level of control based on processes of insight and flexible perspective-taking. These abilities are commonly labeled "metacognitive" to indicate the fact that they operate on top of, and can to some extent control and regulate, other cognitive processes, including emotions.

The emotion researcher James Gross at Stanford University has investigated emotion regulation in some of the greatest detail and has come up with a schematic for how to think about the metacognitive level of control that emotion regulation can exhibit. The figure opposite illustrates his ideas.

Social Communication

There is a very important consequence of the automaticity of emotional behaviors. Since they are difficult to control, they can serve as authentic social signals about an emotion state. You can tell, and other animals can tell, something about the emotion state of a hissing and arching cat, a growling dog, or a crying person. Of course, we can be wrong about the emotion state we infer this way, and most other animals probably do not explicitly infer an emotion state at all but simply detect associations to predict behaviors. Regardless, the point is that emotional behaviors are poised to be co-opted for social communication. Since a hissing behavior originally signaled a high probability of an attack if approached, the cat evolved hissing behaviors also to warn other animals away, as a predictive social communicative signal.

In a given context, and given prior experience, we can deduce emotion states in animals from their emotional behavior—a form of social communication in which we engage automatically all of the time (figure 3.8). But are we right about the emotions we infer? There is a difference between the way other animals perceive emotional behaviors among themselves, the way in which laypeople attribute emotions to animals from their behavior, and the way the scientist attributes emotions to animals and people from their observed behavior. In the first case, there are certainly perceptual mechanisms that detect particular behaviors that have evolved as communicative signals—animals fight, mate, and play in part because they are able to communicate this way. However, they do not attribute emotions to one another, since they don't have the concept of an emotion to begin with.

In the second case, there is some consensus in how laypeople interpret the emotional behaviors of animals, but little evidence that they are generally right. For a few species, especially mammals, there is enough homology in emotional behaviors, and enough experience with observing them, that we are indeed able to communicate with those animals through emotional behaviors. Pet owners know this well. However, we have trouble mapping the behaviors that we observe onto our words for emotion categories. For instance, our cats perform several different kinds of behaviors that are all related to a positively valenced emotional state: they purr, they knead, they roll on their backs, and they vibrate their tails. However, these different behaviors are displayed under different precise circumstances, and it is difficult to know what label to put on the emotion. Some of the behaviors seem to relate to a kind of joyful anticipation, some to social affiliation and cuddling, some to food, some to play. People can infer some coarse emotion properties quite reliably across a large range of different species. For instance, the dimension of arousal (which is not unique to emotions) appears to be attributed by people from

FIGURE 3.8. Expression of emotions in animals. Which of these can you recognize? *Top panel*: There is good consensus for interpreting the expressions of animals with which we have interacted extensively so as to learn their social communicative signals (and for them to learn which ones to express toward us). A dog is cowering in a negatively valenced state that could be fear or submission, and a cat is hissing in a negatively valenced state that could be fear or aggression. *Middle panel*: Many other behaviors are impossible to interpret without more contextual information or ethological training (freezing in fear in a mouse, affiliative hooting expression in a chimpanzee). However, all these expressions can be quantified reliably by careful study, even though we need to be cautious in putting names on them (Parr and Waller 2006). *Bottom panel*: emotional expressions in invertebrates. Despite often very alien behaviors, we can learn to deduce coarse emotion categories even from the behavior of these animals (an octopus fleeing in fear; a male fly displaying wing threat in aggression toward another male fly; see chapter 7).

the vocalizations of many animals, as diverse as elephants, birds, and alligators (Filippi et al. 2017).

The third case, the scientist trying to make sense of animal behaviors tries to disambiguate among these behaviors by incorporating much richer, contextual descriptions. The facial expression, body posture, context, and overall dynamics of the situation and the behavior all need to be taken into account. Ethologists have long tried to make such detailed descriptions of animal behaviors and have generally been cautious in giving them labels like the words that we normally use for emotions. Many of the behaviors reflect expressions of emotion states, but we currently lack the appropriate labels except in a few crude cases. So it is certainly possible to infer something about an emotion state from the observed behavior—but we need to be careful in not anthropomorphizing into human emotion categories (dimensional frameworks may be better in some respects), and we need to interpret the behavior in the situation and context in which it occurs, not in isolation.

It becomes tricky to disentangle the ancestral adaptive role of an emotional behavior from its role as a social communicative signal, and nowhere is this more problematic than with human facial expressions. It is useful to return to the example of clocks we discussed in chapter 2, when we introduced the idea that emotions are fundamentally functional states. There are ancestral functions of emotions, perhaps reflected best in simpler organisms that show some of the building blocks of emotions and not yet all of the properties in figure 3.2. Over time, and with the occupation of different niches for different species, additional building blocks evolved. This simplistic picture of evolutionary change assumes that older functions are retained and new ones are simply added over time. Certainly, that is one thing that happens during evolution.

But another very important mechanism for evolutionary change is to co-opt an existing, ancestral function into a new function. Sometimes that is done by just replacing the original function with a new one. Sometimes it is done by duplicating the original function, so that both the ancestral one and a new, co-opted one can be used. This mechanism is well known to happen at the level of genes, for instance; gene duplication makes it possible for a duplicated gene to mutate toward a novel function, while the original gene (which might be essential) can still carry out its original function.

Facial expressions are probably a mixture of these possibilities. Neural control of facial muscles is complex, and partly volitional and partly automatic. For instance, it is easier to control the lower half of your face volitionally, but harder to control the upper half. This can be seen when people fake a smile: a so-called Duchenne smile is a genuine smile involving both eye and mouth muscles. In the non-Duchenne smile, a person smiles with the bottom half of their face (their mouth), but their eyes don't quite fit and do not show the same expression as with a genuine smile. So we have some volitional control over producing a smile on our face, but there is also some automatic control, and this is produced by a complex mixture of brain regions that innervate the muscles of our face (Müri 2015).

Given the mixed and complex neural control over our face, this may also reflect a mixed and complex set of functions that facial expressions subserve. These range all the way from ancestral functions involved in basic emotion states (for example, baring the teeth if you need to bite somebody to defend yourself), to social functions that are relatively automatic and genuine (for example, baring the teeth to signal that somebody better be careful, otherwise they might get bitten), to social functions that are strategic or deceptive in some species (for example, baring the teeth if you are an actor, or merely to frighten somebody as a prank). If all you are observing is a facial expression, this by itself is ambiguous with respect to what kind of state caused it.

Despite their inherent ambiguity, human facial expressions have been used as the single most common emotion stimulus in all of human psychology research, and probably even in all of human neuroscience research on emotions. In typical experiments, participants are shown a facial expression labeled as fear by the experimenter, and their verbal response or classification of the emotion is recorded, or the response in their brain is measured with fMRI. Needless to say, it is very unclear what such brain responses signify, at least without a lot of further information. It is also problematic to know what our verbal "recognition of emotions" from facial expressions means. Maybe you think the person looks afraid, but I think she is just acting and feels smug. Who is right?

In everyday life, a smile can mean that a person is happy, but often it means they are anxious, submissive, or a host of other socially relevant states for which the smile is used as a signal, often even a strategic and manipulative signal (Rychlowska et al. 2017). There are cultural differences in the meaning of different facial expressions, and complex social rules about when it is appropriate to display those expressions. Depending on the situation, this level of regulation makes it complicated to link a facial expression in humans to an emotion, a topic of considerable debate in psychology (box 3.5).

BOX 3.5. Expression and recognition of facial expressions.

One of the aspects of emotion that has been studied the most intensely, especially in psychological studies of human emotions, is facial expressions. Pioneering studies by Paul Ekman and his colleagues in the 1960s and 1970s suggested that some facial expressions are recognized as expressing the same emotions across all human cultures (Ekman 1972). Ekman traveled to New Guinea and carried out studies in which tribespeople were shown photos of facial expressions, and the emotion category that they associated with the expressions was probed. This led to the idea that there are expressions of a subset of emotions, so-called basic emotions, that are expressed and recognized across all cultures, hypothesized to be based on innate brain modules for their expression.

Ekman's list of basic emotions consists of happiness, surprise, fear, anger, disgust, and sadness, although contempt is also sometimes included and surprise is sometimes dropped. There is some support for what Charles Darwin had called "serviceable associated habits," ancestral functions that facial expressions may have served. For instance, the wide eyes and flared nostrils of a fear expression serve to widen the visual field and maximize the detection of odors, perceptual functions that would have aided the detection of possible threats (Susskind et al. 2008). Recent studies have shown, however, that facial expressions are much more nuanced and complex social signals than mere "expressions" of basic emotions (Fernandez-Dols and Russell 2017).

The ability to recognize emotions from facial expressions, like the ability to perceive faces normally at all, requires experience during development. Abnormal emotional experience, such as that of children who were abused, results in abnormal ability to recognize emotions in facial expressions, especially anger (Pollack and Kistler 2002). Relatedly, cross-cultural studies in humans show that people from different cultures may categorize facial expressions into different concepts, for which they may have different words (Gendron et al. 2014). In general, assessing whether a facial expression (or any other emotion expression) is recognized "accurately," or even just consensually, depends critically on the task; for instance, agreement is much lower when free responses can be given than when a small set of matching labels is provided. In deciding what intensity of facial expression counts as an expression of a given emotion, the brain needs carry out complex pattern completion and pattern separation, and then to set a threshold that separates "correct" instances (true positives) from "incorrect" instances (true negatives). This is essentially a decision problem, which can be formalized using the framework of signal detection theory (Lynn and Feldman Barrett 2014). Different people, and different tasks, can yield different criteria for where to put the boundaries that separate one emotion from another. The accuracy in detecting or discriminating an emotion from a facial expression depends on one's bias and one's sensitivity. For instance, in the terminology of signal-detection theory, high sensitivity and a low bias will result in detecting all possible threat expressions, but also incorrectly detect threat in facial expressions when there is in fact none present (false positives; see box 2.2).

The human face can show expressions through seventeen pairs of facial muscles that we share with great apes (but monkeys have fewer), and the actions of these muscles, and their contribution to a particular expression, can be quantified. The Facial Affect Coding System (FACS) exists for humans, chimpanzees, and macaque monkeys and permits a quantitative description of facial expressions without necessarily attributing any emotion state (which, as noted previously, is culture- and context-specific for humans, and is mostly unknown for other animals). Some quantification also exists for other expressions of emotions, such as those from postures or voice (for instance, the voice can be spectrally decomposed and various attributes can be quantified physically, without referring to emotion categories).

Exactly how the emotions expressed by faces are represented in the brains of observers is complicated (Schirmer and Adolphs 2017). There appear to be representations for discrete emotion categories, for continuous dimensions, and even for individual action units that describe the muscles in the face. There are also neural representations of general social inferences about people's mental states, and all of these varied types of representations are engaged when we look at a facial expression of emotion. This means that the brain can do a lot of different things when seeing a facial expression, all depending on the goals and task at hand. The ability to make inferences about other people's internal states is often called "theory of mind" and engages a network of brain regions involving the temporoparietal junction, dorsomedial prefrontal cortex, and precuneus, among other cortical regions. This neural system, and especially the dorsomedial prefrontal cortex, is activated not only when we attribute emotions to people from looking at their faces but also when we look at the faces of animals to infer how they might feel (see chapter 9).

While emotions do often cause changes in facial expressions, the expression by itself is in general not sufficient to determine the specific emotion. This conclusion is similar to the one we drew in box 2.1: emotions also cannot be uniquely inferred from psychophysiological measures in the body. Nonetheless, our attitudes and social behavior toward other people is strongly influenced by the judgments we make about their faces. Moreover, we often feel strongly confident that we can judge somebody from looking at their face. We're mistaken: our judgments of other people's faces reflect a complex mix of evolved and acquired biases that say more about the stereotypes we hold than about the person whose face we are perceiving (Todorov 2017).

Learning

We already noted that learning is a very important property of emotions, when we discussed generalization. In fact, many of the properties listed in figure 3.2 are both innate and learned. Emotions involve innate components insofar as there are useful regularities that can be detected by sensory systems as inputs to an emotion state. Bitter tastes generally signal toxicity, as do certain chemical compounds that evoke foul smells, and so all of these can serve as unconditioned stimuli to trigger disgust (which does not mean that you are born with the ability to trigger disgust from these stimuli—development is still required, see box 6.1). Looming large shadows overhead generally signal something big falling on you or a predator approaching overhead, and gaping teeth appearing signal you're about to be eaten; these sensory cues can serve as unconditioned stimuli to trigger fear in the same way that expanding dark circles overhead are sufficient to trigger fear in mice (remember figure 2.5). In general, if there is a large expanding shadow overhead and you do nothing, bad things will happen to you rather than good things. In general, if a bunch of teeth are approaching you rapidly, you will get bitten or eaten if you do nothing. Evolution has incorporated these statistical regularities into the innate sensory components of an emotion state, and they correspond to what in the laboratory we would call "unconditioned stimuli" with respect to an emotion.

But of course the range of sensory inputs that can cause an emotion state does not stop there, since there are many other kinds of stimuli that also might signal toxicity or threat, and that should therefore evoke disgust or fear. Those stimuli need to be learned about, and so emotions also involve considerable learning—a key aspect of the flexibility of emotions. As we mentioned earlier, conditioned taste aversion can make you disgusted by what was previously delicious-tasting beer—if you drank so much of it last night that you got sick of it. Pavlovian fear conditioning would be an example of learning with respect to fear: you can come to fear all kinds of things by learning about them. However, even here there are already built-in specializations; it is easier to learn to be afraid of some stimuli than others. For emotions other than fear, learning may be even narrower. It's probably impossible to learn disgust to a tone, since that is just not the sort of stimulus that could plausibly predict basic disgust.

These are aspects of the "domain specificity" of emotions, a term we discussed in more detail in box 3.1. Domain specificity refers to the restricted functional range of stimuli or circumstances that cause an emotion state. Domain specificity is also a key feature that distinguishes emotions from volitional deliberation, which appears not to have any such restrictions on its domain (you can think about anything at all, but you can't become sad or angry about just anything at all). If emotions were completely flexible and only learning was involved, we could treat anxiety and depression in patients just by giving them the right things to learn—there are attempts to do this to some extent, but the difficulty of achieving it shows that the flexibility and plasticity of emotions has limits.

Given the emotion properties we discussed in this chapter, we can next ask whether we find evidence for their presence in the behavior or neural response of an animal or person. To link such data to an understanding of an emotion state, however, we first need to return to the points we made at the beginning of this chapter: how do we know which measures are actually part of the mechanism that is causally responsible for an emotion, as opposed to mere consequences or correlates of the emotion? How do we know which of our brain measures reflect a necessary or a sufficient part of the mechanism for generating an emotion? This requires us to clarify the logic of neuroscience investigations, a topic we take up in the next chapter.

Summary

- This chapter built on the idea that emotion states are functionally defined states. In order to achieve their function, they need to have certain processing features: what they do is determined by how they can do it.
- We provided a preliminary list of emotion properties in figure 3.2.
- Emotions can be situated in a dimensional space and can be related to one another in terms of similarity. A common dimensional space in which to compare emotions is a two-dimensional space consisting of arousal and valence. However, it is currently unclear whether

"arousal" is a unitary property, or whether there are different kinds of arousal for different kinds of emotions (see chapter 5).

- Flexibility and persistence of emotion states allow the integration of multiple sources of information over time. One elegant model that captured this integration was the drift-diffusion model.
- Coordination is a property requiring a particular control architecture. This aspect can be formalized in yet another elegant model: a mathematical graph or network. These examples highlighted the power of identifying properties of emotions that lend themselves well to being captured in computational models.
- Emotions show automaticity, which has historically been juxtaposed to volitional, effortful control of behavior, a dichotomy epitomized in classic dual-process theories such as "System 1" versus "System 2."
- The social communicative aspect of emotions, a feature ubiquitously seen in facial expressions, makes an analysis of their function considerably more complicated.
- Emotions involve both innate and learned aspects.

PART II

Neuroscience

CHAPTER 4

The Logic of Neuroscientific Explanations

The previous three chapters discussed how little we really know about emotions (chapter 1), how to begin thinking about them as functional states (chapter 2), and what general properties or features of emotions one might look for when trying to identify instances of them (chapter 3). To transition to the neuroscience of emotion in the next few chapters, we first need to introduce a few conceptual tools—tools that will allow us to translate the broader concepts and questions of the previous chapters (articulated in the languages of philosophy, psychology, and ethology) into hypotheses that one could test within the framework of neuroscience. This will also require clarifying how conceptual and experimental approaches from cellular and systems neuroscience differ from those in psychology and cognitive neuroscience, a topic we take up again in regard to the specific methods used by these subdisciplines (in chapters 5 and 8, respectively). Recall from chapter 1 that we already stressed that different *approaches* to emotion need not be investigations of different *phenomena*, but instead amount to distinct sets of vocabularies and methods that are used to study one and the same phenomenon (a point strikingly made in figure 1.1, the MRI image of the mother and her baby). Whereas chapters 1 and 2 had mentioned some of the larger differences between the disciplines of neuroscience and psychology, we now focus on the main topic of this book: neuroscience.

Although we believe, and most people agree, that emotions arise in the brain, a biological structure that is the subject of neuroscience, the topic also attracts investigators from many additional fields—psychology, philosophy, psychiatry, computer science, physics, and engineering—that often share little in common in how they investigate their respective domains of study. Given this mix, it is unsurprising that, unlike the case for astronomy, chemistry, or genetics, there is no generally accepted, overarching intellectual framework for understanding emotion. It is like the proverbial blind men groping an elephant in different places and trying to describe the object they are feeling—they all agree that there is something there to explain, but none of them speaks the same language. Despite this diversity of epistemologies, we believe that neuroscience is a core system of investigation and rational explanation for emotional phenomena and functions, and that without it our understanding of emotion will be incomplete.

In the following chapters, we will describe our current state of understanding emotions as seen from the perspective of two different neuroscientific subdisciplines: systems neurobiology and cognitive neuroscience. The former uses a vocabulary grounded in neurophysiology, cellular biology, chemistry, and molecular biology, and eschews talk about the mind. The latter uses a broader vocabulary that necessarily includes terms from psychology, talks a lot about the mind, and typically describes the brain at a macroscopic systems level. It is in particular the essential psychological ingredient, the "cognitive" aspect of cognitive neuroscience, that is the source of substantial difficulties in translating between these two subdisciplines.

Because the mind is patently and closely related to the brain, it is often assumed that neuroscientific explanations of emotion will eventually "map" onto psychological explanations, and vice versa, in a way similar to how biologists have shown that classical genetic explanations of heredity can map onto mechanisms based on molecular biology and biochemistry. But for neuroscience and psychology, this is not necessarily a foregone conclusion. Unlike molecular biology and genetics, there are (at least so far) no scientific laws that provide or necessitate continuity between neuroscientific and psychological explanations of emotion, and the relationship between the two disciplines remains a matter of opinion and debate. There are those who believe that psychological explanations of the mind will eventually be replaced by a physical and biological understanding of the brain, and those who feel that psychology offers a way of understanding mind and behavior that need not depend on or even invoke neuroscience; there are many intermediate views as well. Note that this dichotomy is *not* the same as the classic debate between materialist versus nonmaterialist views of mind versus brain. Both are materialist systems of explanation for natural phenomena, but at very different scales and with different vocabularies. At the moment, therefore, it seems safe to view systems neurobiology, and psychology/cognitive neuroscience, as two different ways of understanding emotions and behavior. That does not mean that one is better than, or will eventually be replaced by, or be reduced to, the other. For now, it just means that they are different schemes for trying to understand one and the same phenomenon—emotions. They may well be complementary and continue to exist as distinct disciplines, making communication between them all the more important.

All that said, it is worthwhile to unpack some of the characteristics and assumptions of these different subdisciplines, so that the general reader can have a clearer idea of what they are, and what they are not, and to dispel any misconceptions that one may have picked up from how these topics are often treated in the popular press. In both fields, there are established and very well respected quantitative methods of investigation, "rules" or "laws" that aim to explain and predict observable phenomena, together with levels of description and units of explanation. In practice, it is also noteworthy that each field sometimes borrows concepts and terms from the other field. This has long been the case in psychology, which rather promiscuously borrows terms that are biological, or even neurobiological, but it is true also in cognitive neuroscience, where, in turn, mental (psychological) functions are often assigned to brain regions, and it is true even in cellular neurobiology where simple psychological functions are also sometimes assigned to circuits, even though no "mind" is typically attributed. As we will explain more at the end of this chapter, and also in chapter 8, this mixing of terms in fact happens frequently in practice but—at least without a lot more explanation—mixing of terms often results in confusion.

Levels of Biological Organization

Neuroscientific studies proceed on the assumption that the mechanisms that produce emotions, just like those that produce perception, memory, or attention, are realized in functions of the brain and will ultimately be understood in terms of the brain's underlying biological, physical, and chemical processes. However, brain functions can be studied at many different "levels," or physical scales. One way of viewing levels parallels the different, nested scales of biological organization that you can see as you use progressively more powerful microscopes: at a coarse scale, you see cells and their nuclei, but not much of what is inside them; at more detailed (granular) levels, which require higher magnification, you see new structures and new organization (for example, ribosomes and mitochondria) that were not visible at the coarser levels. However, this more microscopic level of detail is not something separate from the more macroscopic level. Instead, the microscopic levels constitute the more macroscopic level, but in terms of different objects or structures.

This common usage of "levels" in neurobiology, often referred to as "levels of description," then, refers to scales of biological organization: molecules, synapses, cells, local circuits, large-scale distributed networks, and entire brains (figure 4.1). A neurobiologist investigating emotion circuits in the hypothalamus in a mouse might be focused primarily at the level of circuits composed of populations of neurons, whereas a cognitive neuroscientist using fMRI to investigate emotions in humans might be focused primarily at the level of large-scale networks distributed over the entire brain (figure 4.1). There is far from universal agreement as to which of these levels might be the most appropriate to understand a given brain function—and perhaps different levels are best suited for investigating different brain functions. Each level has its own units of explanation and commonly accepted definition of "mechanism," which we will discuss further.

Some neuroscientists believe that it will be possible, eventually, to "bridge" different scales; that is, to explain brain mechanisms at a coarser level of description in terms of processes at a more granular level, like a set of nested Russian dolls. However, that is currently an aspirational goal; there are some neuroscientists who believe it is a hopeless one, because the brain functions they are interested in—such as object recognition by the visual system—are "emergent properties" of complex systems comprised of many interacting components, which

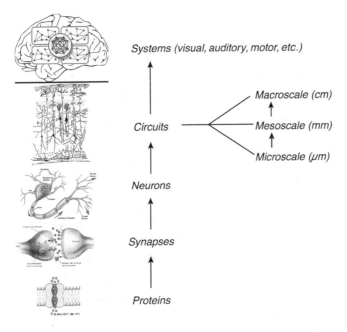

FIGURE 4.1. Different scales of biological organization in the brain. *Proteins* (e.g., ion channel, neuropeptide) are a microscopic unit of biological organization. Red spheres indicate ions (e.g., sodium) moving through a protein pore in the membrane of a neuron (blue), a process that generates the electrical potentials by which neurons communicate. *Synapses* (and small structures within a cell) are composed of hundreds or thousands of different proteins, carbohydrates, and lipids. Red spheres here indicate individual neurotransmitter molecules (e.g., glutamate) released from the axon terminal (*left*) that bind to receptors on the dendrite (*right*), enabling communication between two cells; drawing not to scale. *Neuron* illustrates the level of a single, canonical neuronal cell, which may make and receive hundreds or thousands of synapses with other neurons. *Circuits* comprise connections between hundreds, thousands, or even millions of neurons, and vary across length scales from microns (µm) to centimeters (cm), depending on the size of the brain and whether they are local (microscale) or distributed (mesoscale, macroscale). Systems comprise multiple circuits that implement particular processes or functions, e.g., different types of sensory systems or motor systems.

cannot easily be predicted from the study of the individual components at lower levels. Again, there is agreement that all physical constituents of the more macroscopic level are contained in the more microscopic level—brains are clearly made up of cells and their connections, and there is no magic additional ingredient at the more macroscopic level— but people disagree whether we can understand, explain, and predict the phenomena characterized at the more macroscopic level in terms of the phenomena characterized at the more microscopic level. As a simple example, consider "wetness" of water, the fact that it sticks to and coats many materials. Wetness is a bulk property, which you seem to lose if you look at single molecules of H_2O. Wetness just isn't there at the level of single molecules; rather, it emerges as a higher-level property (that is, nonetheless, constituted entirely by a bunch of interacting water molecules—there is no additional magic ingredient added).

An example from neuroscience analogous to the "wetness" of water example might be a large-scale network property. For instance, as we will see in chapter 8, many fMRI studies of cognitive function in humans now look at the properties of whole-brain networks that are defined by correlations observed between the network components. Cognitive neuroscientists thus talk about "hubs," "small-world organization," and "network centrality." These are similar to network properties that can be used to describe the internet. But they do not find any mapping to a cellular or microcircuit level of description. So, for example, a scientific study describing a finding on attention or memory as a change in network centrality, from fMRI studies in humans, would be difficult to relate to a description at the cellular level, even though its results of course ultimately have to be constituted by cellular-level phenomena.

In the case of emotion, we just don't know the best level at which to understand it; therefore, neuroscientists are pursuing its study at multiple biological scales, all in parallel. Essentially all of the different colored regions of spatiotemporal resolution relevant to monitoring and manipulating brain activity shown in figure 4.2 are represented in modern neurobiological studies of emotion.

The Concept of Mechanism in Neuroscience

Throughout this book, we often characterize the relationship between brain processes and emotion states using several interchangeable words. We say that brain events cause, produce, generate, implement, are involved in or realize emotion states. We take all of these as consistent (although not equivalent)—they emphasize different aspects of the relation between brain and emotion, and differ in the degree of experimental rigor with which those relationships have been established.

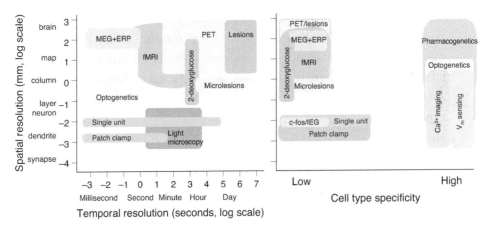

FIGURE 4.2. Methods for investigating the brain at different scales. Some of these measure brain activity directly (single-unit recordings) or indirectly (fMRI; see chapter 8), whereas others can manipulate brain activity (optogenetics, lesions). There are more techniques than just the ones listed here, and of the ones listed, there are many subcategories one could describe. This figure is intended just to give a broad overview, and to make the point that a huge range of spatial and temporal scale is covered by current neurobiological methods. *Left*: plotting methods in terms of their resolution in space and time. *Right*: plotting methods in terms of their cell-type specificity. Note that some very microscopic methods, such as single-unit extracellular recordings, do not have any cell-type specificity (we can record from a single cell, but we generally cannot select or even know which type of cell is being recorded). Abbreviations: MEG: magnetoencephalography; ERP: event-related potentials; fMRI: functional magnetic resonance imaging; PET: positron emission tomography; c-fos: a gene whose expression is induced when a cell is active, serving as a cellular label of activity (see box 6.3); IEG: immediate early genes, category including c-fos but also other IEGs such as Arc; Ca²⁺ imaging: detection of neuronal activity using a calcium-dependent fluorescent indicator such as GCaMP6; Vm: membrane-voltage sensing using fluorescent indicators.

For most scientists, the ultimate goal of scientific explanation and understanding is a complete description of a causal mechanism—a description so complete that it would allow us to predict the behavior of a system from the description, would allow us to intervene and manipulate the system to produce specific results, and could even allow us to build such a system from scratch, at least in principle. By "mechanism" we thus mean an account of a neural process or function in terms of units at a certain level of granularity (biological scale), and the causal relationships between these units. The kind of picture we have in mind when we talk about mechanisms is similar to what we showed in figure 3.5 in the previous chapter—some scheme for understanding how different components of a system influence each other in a causal chain of events, to achieve a particular function. For example, a mechanistic explanation for communication between neurons at the cellular/physiological level would be that action potentials (electrical signals that travel along an axon) when they arrive at the nerve terminal *cause* an influx of calcium, which in turn *causes* synaptic vesicles (small packets of neurotransmitters such as acetylcholine) to fuse with the surface membrane, releasing their contents into the synapse where they bind to receptors on the surface of the postsynaptic neuron, *causing* electrical changes that are converted into yet more action potentials, and so on.

Mechanisms can be described at any of the different levels of description we alluded to previously. For instance, a molecular-level explanation for the process outlined in the previous paragraph would describe it in terms of the proteins in the membrane of the neuron that *cause* the action potential to occur (for example, voltage-gated sodium and potassium channels), the proteins in the nerve terminal that *cause* calcium entry (voltage-gated calcium channels), the proteins in the synaptic vesicle and nerve terminal that *cause* fusion in response to calcium influx, and the receptor proteins in the postsynaptic neuron that bind the neurotransmitter and convert that binding into an electrical signal. Nobel Prizes have been awarded for the research that produced a mechanistic understanding of nerve communication at each of these different levels. Importantly, while the *units of explanation* at the physiological level are electrical currents flowing across the neuronal membrane, and the chemicals released by the neuron at the synapse, this mechanism can be recast in terms of more fine-grained units of explanation: proteins, lipids, and other molecular components of the neuronal membrane and cytoskeleton. It is also possible to explain the molecular mechanism in terms of atomic mechanisms, through structural studies of neuronal proteins and their interactions, and Nobel Prizes have been awarded for that as well. At each level, "mechanism" involves a description of *causal chains of events* involving interactions between "units of explanation"—electrical currents, molecules, atoms—and the consequences of these events; in other words, the what, the when, and the how. It is a story science tells about how a process unfolds, with characters writ large or small.

Correlation versus Causation in Neuroscience

In genetics, molecular biology, and more recently neuroscience, mechanism implies causality: a mechanistic understanding of a given process must incorporate a description of causal relationships. By "causality," we mean that if A causes B (A→B), then B happens as a consequence of A. Such a causal relationship will result in a correlation between A and B, in observational data (figure 4.3, 1a). But the converse is not true: just because two things are correlated (in space or in time), does not mean there is a causal relationship between them (figure 4.3, 1b–d).

For example, in human neuroimaging studies, it is often mistakenly assumed (especially in the popular press) that correlations between brain activity, and a given behavior or emotion state, imply that the brain activity causes the behavior or state. Thus, for example, if activity in the human amygdala is, under certain circumstances, correlated with the subjective experience of fear, one might be tempted to infer that activity in the amygdala *causes* fear. But this would be logically incorrect: *correlation does not imply causation.* It is equally possible that the experience of fear causes activity in the amygdala, or that some other unobserved process causes *both* increased amygdala activity *and* the experience of fear at the same time (figure 4.3, 1b–d); in the latter case, we would say that there is *no* direct causal relationship between amygdala activity and the experience of fear. While computational methods are being developed to try to extract causal inferences (defined more broadly) from correlative neuroimaging data (see chapter 8), the most direct method to distinguish correlation from causation is through interventional tests of *necessity* and *sufficiency.*

We say that "A is necessary for B" if experimentally blocking process A prevents process B from occurring, and that "A is sufficient for B" if experimentally activating process A causes process B to occur outside of its normal time or place, or more strongly. Ideally, both necessity and sufficiency should be demonstrated in order to conclude that, for example, gene (or cell) "A" causes process or behavior "B" to occur ("A→B"; figure 4.3, 2a). The reason that both are important is because necessity alone does not distinguish whether an element in a hypothesized causal pathway (for example, "A") plays a "permissive" or "instructive" role.

1. Correlation vs. causation

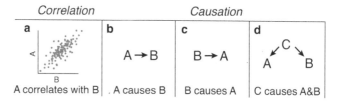

2. Causation: necessity vs. sufficiency

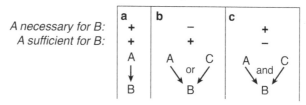

FIGURE 4.3. (1). Correlation versus causation. (1a) The activation of a gene, cell, or process "A" is correlated with process "B." This is indicated by the data one typically measures; in this simple example, the relationship between A and B is linear, with some measurement noise, producing a scatterplot that lies along a straight line. Whenever A is small, B is also small; whenever A is large, B is also large. (1b–1c) Three types of causal relationships between "A" and "B" that are all compatible with the observed correlation plot in (1a): (1b) "A" causes "B." (1c) "B" causes "A." (1d) "A" and "B" have no causal relationship to each other, but are both caused independently by gene/cell/process "C." (2). Necessity versus sufficiency. In (2a–c) gene/cell "A" causes process "B"; however, in (2a) gene/cell "A" is both necessary *and* sufficient for process "B" to occur. In (2b) either gene/cell "A" *or* gene/cell "C" alone can cause process "B"; therefore "A" is sufficient, but not necessary, as long as gene/cell C is operational. (2b) Genes/cells "A" *and* "C" are both required for process "B" to occur. Therefore gene/cell "A" is necessary, but not sufficient to cause process "B" in the absence of gene/cell "C". Note that the arrows in these diagrams do not represent direct interactions, like inputs to a device in an electrical circuit (e.g., "and" gates versus "or" gates); rather, they symbolize the *consequences* of the activation of gene/cell "A" for the occurrence of process "B" (e.g., a measured behavior).

This crucial difference is best appreciated by analogy to an automobile: gasoline is "permissive" (or "enabling"), in the sense that if you drain the gas tank, the car will not run. However, it is not "instructive," in the sense that a car with a full tank does not go any faster than a car with a half tank of gas. In contrast, the accelerator pedal is "instructive"; if you press harder on it, the car goes faster, *and* the car will not run at all without it. In other words, the gas pedal is both necessary *and* sufficient to control the speed of the car, but it requires the enabling (permissive) factor of a full gas tank. Similarly, in the brain, a region necessary for fear could be permissive in the sense that it is required to process the sensory stimuli that trigger fear (for example, the cochlear nucleus for a fear-evoking auditory stimulus), but that does not mean that this activation is *sufficient* to cause fear. Demonstrating sufficiency requires a *gain-of-function* intervention, like activating a particular brain region, and cannot be inferred from a *loss-of-function* manipulation alone (for example, a lesion).

Importantly, while "sufficiency" can distinguish permissive from instructive effects, it does not necessarily imply a *direct* causal influence. For example, the demonstration that optogenetic stimulation of a given brain region or cell type is "sufficient" to promote a particular behavior or state does not mean that the stimulated neurons directly produce that effect: there may be many unidentified intervening steps or circuit nodes. As we shall see in chapter 6, circuits involved in emotion states are situated in what one might call the "inner brain," regions located midway between circuits that process sensory input, and those that organize motor output. The activation of neurons in such a region introduces a pulse of activity into a complex circuit that ramifies and spreads throughout the brain, as indicated by a few studies that have monitored the effect of optogenetic stimulation of the rodent brain using fMRI (see figure 11.2). Discovering the mechanism by which such stimulation produces its observed behavioral effects requires extensive follow-up work, and is crucial to the interpretation of such studies.

Note that there may be situations in which gene/cell A is sufficient but not necessary (figure 4.3, 2b), or necessary but not sufficient (figure 4.3, 2c), for process B. Sufficiency without necessity can obtain if there are redundant processes that promote process B, in addition to gene/cell A; those redundant pathways can compensate if gene/cell A is eliminated, therefore "A" is not strictly necessary (figure 4.3, 2b; "C"). Conversely, gene/cell A may be necessary, but not sufficient, because process B requires gene/cell A plus one or more additional type of gene or cell ("C" in figure 4.3, 2c). Therefore, simply activating gene/cell A will not suffice to promote process B, if gene/cell C is not concurrently activated. Thus, a demonstration of either necessity or sufficiency is in itself important evidence of causality.

Note that we are using "causality" here in a deterministic (whenever A happens, it causes B), rather than in a probabilistic sense (if A happens, there is an increased chance that B will happen). An example of the latter is the oft-quoted statement that "smoking cigarettes *causes* lung cancer"; in our usage, this statement is inaccurate. Smoking cigarettes increases the *chances* that you will get lung cancer, but not everyone who smokes cigarettes will get lung cancer, and some people will get lung cancer even if they never smoked. Therefore, smoking cigarettes is a *risk factor* for lung cancer, in a statistical or epidemiological sense, but it is neither necessary nor sufficient for lung cancer in all people.

Testing Causal Relationships between Neural Activity and Behavior

Having an experimental logic to test for necessity and sufficiency is one thing; actually having the tools to do it is another. Advances in molecular biology have provided extremely precise tools to manipulate specific genes. Using these tools in genetic "model organisms" such as mice, fruit flies, and roundworms, it has been possible to test the necessity and sufficiency of individual genes for particular behaviors by "knocking them out" (that is, mutating them to inactivity) or overexpressing them, respectively. While such experiments can provide evidence of a "causal" relationship between a gene's function and a behavior, such observations do not yet explain *how* the gene influences the behavior, mechanistically; for example, the gene could act during embryonic development to influence the formation or wiring of a brain region necessary for that behavior in adulthood, or it could act in the adult to influence activity in the circuit. Without knowing what the circuit is, however, it is difficult to distinguish these hypotheses.

This realization led, in the early 2000s, to a push to develop methods to test the necessity and sufficiency of specific groups of *neurons* in experimental animals. In this case, rather than inactivating or activating specific *genes*, methods were developed to inactivate or activate specific *cells*, in specific brain regions at specific times. For example, an approach called "optogenetics" (see box 5.1) uses light to turn specific classes of neurons on or off, with millisecond time resolution. Note that the region for "optogenetics" in figure 4.2 is spatially more macroscopic than single-unit recordings. This is because optogenetics (as most commonly applied) does not manipulate a single neuron, but rather a genetically identified *population* of neurons—often hundreds or thousands of cells or more. Activating or inhibiting a genetically defined population of neurons, as with optogenetics (boxes 5.1 and 5.3), can have a profound effect on a specific behavior. In flies and worms, activating a *single* neuron can sometimes strongly influence behavior as well, but this is rarely the case in vertebrates. These methods, and other related techniques, have revolutionized our ability to assess causal relationships between the activity of particular classes of neurons and specific brain functions.

Interventional tests of necessity and sufficiency in the brain typically require invasive experiments; one needs to literally insert tools for manipulating neuronal activity into the brain. While such experiments are routine in some animals (especially rodents, as described in chapter 5), in humans they are technically challenging and often ethically impermissible. Instead, one has to rely on accidents of nature, such as the case of the patient known as S.M., who had focal lesions of the amygdala, and whom we describe in box 8.2. But such accidents of nature or medicine remain rare and difficult to reproduce. Other techniques, such as transcranial magnetic stimulation, can be applied in healthy people to cause temporary functional lesions in prespecified locations—but are not complete and have very poor anatomical resolution. It is also possible to activate regions of the brain experimentally, for instance, through deep-brain stimulation. This approach is in fact being used to treat disorders of emotion, such as depression. However, once again it is poorly controlled, and it can be applied to only a small subset of people with strong clinical justification. The different methods typically used to study emotion in animals versus humans have pervasive effects on all aspects of the neuroscience of emotion. We will return to a more detailed description of specific methods in boxes 5.1–5.5 (for optogenetics, calcium imaging, and related technologies) and chapter 8 (for fMRI).

Levels of Abstraction

We discussed levels of biological organization earlier in this chapter, when we described going from more macroscopic to more microscopic ways of describing the brain (figure 4.1). There is another usage of "levels," which we will call *levels of abstraction,* an approach the neuroscientist David Marr had first described (figure 4.4) (Marr 1982). We have mentioned this already, but provide a little more detail here. According to Marr, there are three broad levels of abstraction (what he called "levels of analysis"): the *functional or ecological level,* which explains the abstract functional problem that the organism needs to solve (for example, avoid being eaten by a predator; Marr somewhat confusingly called this the "computational" level); the *algorithmic or computational level,* which explains the algorithms and computations for achieving this function (for example, run if there is no place to hide; hide if there is a nearby hiding place—as done by the mouse in the experiment we showed in figure 2.5); and the *implementation or neurobiological level,* which explains how such algorithms could in fact be realized in the neural circuitry of the brain. Of course, one could easily interpolate many intermediate levels into this scheme. For instance, the algorithmic level is not a single level, but instead ranges from very coarse conceptual boxes-and-arrows models seen in some psychological theories, to very specific and detailed computational models with mathematical formulas, or even scripts of computer code that can be executed in a simulation. In general, the functional level is compatible with many different algorithms for achieving the same goal, and a given algorithm can typically be implemented in many different ways in neural hardware. Consequently, one would expect that an abstract, functional-level, or algorithmic-level analysis of an emotion is compatible with many different ways of realizing this in the brains of different species—or even within the brain of one species. In this respect, the *levels of abstraction* scheme differs from the *levels of biological organization* scheme we discussed earlier in this chapter (figure 4.1). All of the different levels of biological organization shown in figure 4.1 are compatible with the "neurobiological" implementation level shown in figure 4.4. Each higher level in figure 4.4 is compatible with multiple, different lower levels (a given functional problem can be solved with multiple algorithms, which can in turn be implemented in a variety of ways).

We should comment on one aspect of figure 4.4 that has now become ubiquitous in both neuroscience and psychology: the metaphor

Level	Discipline	Questions
Ecological	Comparative ethology	What problems is this emotion adapted for?
Computational	Psychology	What algorithms solve these problems?
Neurobiological	Neuroscience	What neural mechanisms implement these?

FIGURE 4.4. Levels of abstraction for investigating emotions. The leftmost entries describe the level of analysis used by the scientist, which ranges from high-level functional analyses of an emotion state (such as used by ethologists or evolutionary psychologists), to computational analyses that draw on the properties required by emotion states to achieve these functions (like the list of properties we provided in figure 3.2), to the investigation of how this is actually generated by the brain (the implementation level). Each level of abstraction asks its own kind of questions, and the questions asked at the different levels often give different degrees of satisfaction to investigators from different disciplines: an ethologist finds the top-level question interesting and the lower levels often irrelevant, whereas a circuit neurobiologist may find the lowest-level question the most relevant.

of the brain as a computer. When we speak of "computations" and "algorithms," there is the underlying assumption that nervous systems compute in some sense. This is a strong assumption, and there are certainly people who disagree with it—we give some quotes in box 4.1. There is also a related popular view that brain computations are reserved for "cognitive" functions—decision-making, problem solving, strategizing, learning, and memory—while the "emotional" part of the brain is essentially chemical in nature, comprising a "soup" of neuromodulators such as serotonin, dopamine, norepinephrine, and endorphins that bathes the brain. This dichotomous view is based in part on the serendipitous discovery that certain drugs with powerful effects on mood and emotion, like Prozac, act by elevating the level of certain chemical modulators (in this case serotonin), throughout the brain. But these drugs are not specific to emotion systems (indeed, they can cause many unwanted and paradoxical side effects); rather, it is the neural circuitry, and not the chemistry, that defines the emotional effect of taking a drug. Thus the popular dichotomy between the "electronic" cognitive brain and the "chemical" emotional brain is a false one. Brain circuit function—described as computation—is influenced by its chemistry, and brain chemicals exert their effects by altering circuit computations. What remains to be understood is exactly how and where in the brain different chemicals act on brain circuits to modify their function, and how that impacts both cognition and emotion. In our view, emotions can be analyzed at an algorithmic level, described as computations, just like any other brain function.

BOX 4.1. Is the brain a computer?

Because the idea that emotions could be described as computations sounds intuitively wrong to some people, we want to say a few more words about this and explain a little more why we believe this reservation is misplaced. Some well-known scientists have voiced this objection. For instance, writing in the *Huffington Post*, the neuroscientist Antonio Damasio says,

> living organisms, including human organisms, use code-dependent algorithms such as the genetic machinery. But while, to a certain extent, living organisms are constructed according to algorithms, they are not algorithms themselves. They are consequences of the engagement of algorithms. The critical issue, however, is that living organisms are collections of tissues, organs and systems within which every component cell is a vulnerable living entity made of proteins, lipids and sugars. They are not lines of code. (http://www.huffingtonpost.com/antonio-damasio/algorithmic-human-life_b_10699712.html)

Damasio's piece is provocatively titled "We Must Not Accept an Algorithmic Account of Human Life." But armed with the discussion of this chapter, you should be able to see why this quote is not an argument against a computational level of analysis of life. Organisms and cells are not computer programs any more than they are molecules; they are a particular organization of molecules that, from a certain vantage point of the scientist, implement computations that all coordinate toward particular functions ("life" in the broadest sense of Damasio's quote).

There are many other examples of the view above. The psychologist Robert Epstein penned a piece in *Aeon* titled, "Your Brain Does Not Process Information, Retrieve Knowledge, or Store Memories. In Short: Your Brain Is Not a Computer" (https://aeon.co/essays/your-brain-does-not-process-information-and-it-is-not-a-computer). The philosopher John Searle is famous for his arguments that the mind does not arise from computations or algorithms (Searle 1980). To see the intuition behind many of these arguments, consider running a very detailed simulation of the weather on a supercomputer. Make it as detailed as you like, modeling every individual atom in principle; still, nobody would actually get wet with rain. John Searle's point is along the same lines: the causal properties of the brain amount to more than implementing algorithms; according to Searle, the brain causes the mind in the same way the stomach causes digestion of food, and merely running an algorithm of either doesn't generate a mind or digest anything.

These viewpoints are important because they force us to clarify what we mean by a "functional" account of emotion states that we described in chapters 2 and 3. We don't mean that emotion states are only functional states, or only computations, any more than that they are only brain states. Remember what we wrote about figure 1.1 a while back, the figure showing the MRI of the mother and her baby: there, we made the point that emotions could be described from quite different perspectives that are not incompatible. The functional level describes what emotions do in the most abstract terms, the algorithmic level describes how they do it more formally (typically, in terms of computations), and the neurobiological level describes how this is realized in a physical substrate such as the brain (figure 4.4). The mere formulation of a function or computation by itself—as in writing an equation on a blackboard—is of course causally completely inert and does nothing. The functional and algorithmic levels should be thought of as abstractions—abstractions provided by the mind

of the scientist—that can help us to understand, explain, and manipulate emotion states. The bottom line for us is that all of these
different levels of analysis are required in a science of emotion.
None can be omitted, and the challenge is how to relate them to
one another.

Mixing of Terms in Neuroscience Explanations

In our previous discussion of neurobiological concepts, we highlighted
the idea of a mechanism. Usage of the term "mechanism" in psychology
and cognitive neuroscience has the same flavor as in our discussion of cellular and systems neurobiology—it also critically revolves around causality, but in this case causal relations between states that are at least in part
psychologically defined. "Mechanism" in psychology typically amounts
to a *hypothesis* about a chain of events, which may be described entirely
in psychological terms or may be a mixture of biologically, behaviorally,
and psychologically defined states. Thus, describing a feeling of fear as
resulting from bodily expressions of fear, which in turn result from perceiving a fear-inducing stimulus, amounts to a psychological mechanism
for how we feel afraid. This particular example in fact paraphrases the
hypothesis put forth by the psychologist William James that we discussed
in chapter 2. A competing hypothesis would be that perception of a fear-
inducing stimulus first causes us to feel afraid, which then causes bodily
expressions of fear. It is clear that many intervening causal steps have been
left out in this description. Cognitive neuroscience typically includes
some mixture of such psychological hypotheses but puts neurobiological
structures, systems, or networks into the components in order to fill out
the mechanism. So, a cognitive neuroscience version of William James's
original hypothesis might propose that seeing a fear-inducing stimulus
causes activation of the amygdala or hypothalamus, which in turn causes
changes in the body, which are sensed by the insula, which make you feel
afraid. We will sketch out more detailed modern versions of precisely this
hypothesis in chapter 10.

As seen in this example, the descriptions of psychological and neuroscientific mechanisms often blend into each other. For example, brain

regions, or systems of them, are sometimes associated with certain psychological functions, and so some descriptions might put "amygdala" into a box within a psychological mechanism based on boxes and arrows. Conversely, it is common in cognitive neuroscience to see discussions about activation in particular brain regions that assign psychological functions to those regions, again providing a mix of neurobiological and psychological terms in postulating a mechanism. Such a mix is fairly common not only in the investigation of emotion in humans but also in many animal studies of emotion. These mixed explanations can be problematic—not so much because they simply use terms from two different disciplines, but because, implicitly or explicitly, the mixture is used to suggest some kind of deeper explanation. Psychological studies that put "amygdala" into a box for "fear" are suggesting that we have an explanation now for how a psychological process might be biologically grounded. Cognitive neuroscientists who discuss their finding of amygdala activation in an fMRI study as "increased fear" are similarly suggesting that they have found a link between brain activations and a psychological process. These examples could be warranted, if the study itself actually demonstrated such a link. But they are often provided based on data too weak to sustain such a claim, or on background assumptions derived from other work.

In cognitive neuroscience, the assignment of psychological functions to activations in particular brain regions is called "reverse inference" and is problematic in many cases (Poldrack 2006). Suppose that several fMRI studies find that the amygdala is activated whenever people report feeling fear. I now show people a variety of pictures and the only thing I measure is brain activation. It turns out that some pictures activate the amygdala more than other pictures. I conclude that these pictures must therefore cause people to feel fear. The well-known problem with this argument is that (a) many other brain regions might also be activated more by these pictures, and (b) the amygdala might be activated by many more processes than the experience of fear. This faulty logic should be familiar to you by now—we began chapter 2 with a list of such problems. Note that it certainly can be possible to infer psychological states from brain activations—but you need a large amount of data from many studies to make this statistically convincing and to rule out confounding possibilities. For f MRI studies in humans, the number of studies is indeed becoming huge—there are thousands of studies on emotion by now—and there are in fact tools available to allow researchers to make such inferences and to put estimates of likelihood on them (Poldrack 2011). If you have a hundred studies showing that the amygdala is activated when people report feeling afraid, and a thousand studies showing that the amygdala is not activated by all kinds of psychological processes other than fear, then you might be justified in inferring fear from measuring amygdala activation. We'll return to this issue in figure 8.3 with an example.

Just as neurobiological mechanisms come in different degrees of granularity that correspond to different levels of description, from systems level to microphysical, cognitive mechanisms also come in different degrees of granularity, corresponding to part-whole relationships between psychological processes. There are very coarse mechanisms illustrated with boxes and arrows, and there are fine-grained models that allow fitting of parameters in an equation that describes the dynamics of a biophysical process. As with the neurobiological mechanisms, where more microscopic is often thought to replace more macroscopic, a more detailed computational model in cognitive neuroscience is often thought to replace a broader conceptual model. This is fairly unsurprising; in both cases, we are simply saying that providing more detail will just absorb a mechanistic explanation with less detail. In the case of neurobiology, the levels of biological organization that framed figures 4.1 and 4.2 were fairly concrete and had little disagreement: it just corresponds to what you can measure with different tools, from an MRI down to a light microscope down to an electron microscope. In the case of cognitive neuroscience and psychology, the granularity of mechanistic descriptions is much less clear, and there is little agreement. One attempt to provide an inventory of terms used by these disciplines and how they are related is given in the "cognitive atlas" (www.cognitiveatlas.org). A quick perusal of this site will give you an idea of how much work there still is to be done in coming up with a list of psychological terms that smoothly relate to one another, and that, taken together, provide a multiresolution description of the mind at all levels. We really have nothing like this at all yet. It is in part for this reason that our book focuses on the neurobiology, rather than on the psychology, of emotion.

Necessity, Sufficiency, and Normalcy

Establishing the necessity and sufficiency of a particular group of neurons for a particular brain function is itself necessary, but not sufficient, for achieving an understanding of how that function or behavior is mediated by the brain. For example, if one considers a simplistic view of a neural circuit as a pathway that detects a sensory stimulus, and relays it via several intermediate nodes to a behavioral (motor) output, then multiple nodes in such a relay might prove to be "necessary and sufficient" for the behavior, as tested by techniques such as optogenetics or pharmacogenetics (box 5.1). But that does not mean that every node has the same function in controlling the behavior; if that were the case, then why should there be multiple nodes? Optogenetics, despite its power and specificity, can be a relatively blunt instrument that, when used to stimulate particular groups of neurons, imposes artificial patterns of activity on these cells and indirectly increases the activity of many interconnected brain regions.

In order to understand the actual function of these cells and their associated networks, it is necessary to study the *normal* firing pattern of the neurons during whatever function or behavior the brain is carrying out. Traditionally, this is accomplished by measuring electrical signals in the brain using microelectrodes. However, it can be challenging to identify the specific cell types responsible for the measured signals. New techniques, based on optical imaging using genetically encoded sensors of calcium or membrane voltage, are proving equally revolutionary as optogenetics in providing a literal window into the brain while it is carrying out its various functions in freely behaving animals (box 5.2). Optical imaging gives us a dynamic movie of the activity of a population of neurons in real time. Furthermore, optical sensors can be genetically addressed to specific neuronal cell types. For many central processes small, genetically defined, populations of neurons ("cell types") may be the fundamental unit for understanding brain function. This would be a level of description larger than a single neuron, but smaller than entire brain structures (as anatomy usually defines such "structures," for example, the amygdala). Again, these tools are invasive and cannot be broadly applied in humans. But their combined application in animal models affords an approach to understanding brain mechanisms underlying emotion and other internal states that encompasses both quantitative correlational measurements *and* tests of causality. In the future, once we have characterized the normal firing patterns actually evoked by a real stimulus, it should be possible to reinstantiate this same pattern artificially using optogenetic stimulation. In this way, we may be able to literally "record" and "play back" specific spatiotemporal patterns of neuronal activity and determine their effect on behavior and internal states, something that has already been achieved for some aspects of memory. Watching neuronal dynamics and perturbing these dynamics in real time promises new insights into brain function that were unthinkable even a decade ago.

There is a final important consideration on which we want to close, related to the discussion of "normalcy" we had previously. Each person's brain is unique. What if your brain and my brain are literally wired differently, so that whatever conclusions you draw about how an emotion is caused in your brain may not apply to my brain? This is a question about individual differences, a topic that psychologists and psychiatrists have spent a lot of time investigating, but which is relatively new to cellular and molecular neurobiology. Although it is less acute a problem in experiments involving genetically inbred strains of animals (as commonly is the case in studies with mice, for example), even in those cases, no two animals will behave in exactly the same way, and no two brains are exactly identical. In principle, differences between individual brains could make our investigation of emotions extremely difficult. Everything we have written so far assumes that two people's brains are sufficiently similar that we can generalize across people in terms of how emotions are generated by the brain, at least to a large extent.

Empirically, while there is indeed individual variation, it appears so far that different brains within a species—and indeed, brains across related species—share sufficient similarities that we can derive generalizable principles for how emotions are implemented across all the different levels we mentioned earlier. In particular, both levels of mechanism and levels of abstraction can help here. It may be the case that the precise wiring of synapses in my amygdala and in yours are different, but that this microscopic difference would disappear if we use a spatially coarse measure, such as the activation seen with fMRI. Similarly, although a mouse brain and a human brain are very different in many ways (for example, circa 100 million neurons versus 80 billion neurons, respectively), a functional description of fear may look quite similar in the two species, permitting some comparisons. Many of the general properties of emotions we listed in figure 3.2, for instance, seem to hold across different people and different species. But individual differences do of course exist, and they emphasize one critical methodological point: we need to be very clear about the organism we are studying, and, in human studies, we even need to be very clear about the demographic and cultural background of our subjects.

Armed with the conceptual tools of this chapter, we will next turn to discussing the neurobiology of emotion. This will involve highlighting particular examples that are best understood, and it will draw on specific tools that allow us to make inferences about mechanisms—about necessity, sufficiency, causality, or correlation. Chapters 5, 6, and 7 detail examples from studies in animals, where we have by far the greatest degree of experimental control, and where we have the most detailed, most microscopic descriptions available. Chapters 8 and 9 then survey some studies in humans, which are much more macroscopic, and which make stronger links to psychological terms. One of the most important challenges in the science of emotion is to provide some continuity or bridge between these two sets of investigations—for instance, by applying the tools currently used in animal studies also to studies in humans, or conversely. We close the book with a list of suggestions for possible future experiments that might begin to provide such a bridge—these are not experiments currently being done either in animals or in humans, but they are the experiments the next generation of emotion scientists should begin to tackle.

Summary

- Both neurobiology and psychology are needed to constitute a science of emotion, but these two disciplines often use different terms and approaches. The main two subdisciplines of neuroscience that will be the focus of this book are systems neurobiology (the level most commonly used to investigate emotions in animals) and cognitive neuroscience (the level most commonly used to investigate emotions in humans).

- A central concept in neuroscience studies of emotion is the notion of a causal mechanism. Correlation does not imply causation. Molecular, cellular and systems neuroscience use interventional experimental methods that test necessity and sufficiency, in order to determine whether a correlation between neuronal activity and a behavior or state reflects a causal relationship between the former and the latter. Cognitive neuroscience is developing computational tools to extract causal inferences from correlative data in cases where interventional experiments are difficult to perform, such as in humans (chapter 8).

- Investigations in neuroscience take place at different levels of description, that is, scales of biological organization (figure 4.1). As we noted in figure 4.2, there are many different tools used by the neurobiologist, and different tools have different resolution in space and time.

- Systems neuroscience has benefited from an explosion in the development of genetically based tools for marking, mapping, and manipulating neural circuits, including optogenetics, chemogenetics, and calcium imaging. These tools are useful in genetic model organisms, such as worms, flies, and mice, but are currently difficult to apply in genetically less tractable systems, such as rats and monkeys—although that is gradually changing. Research and therapeutic application of these tools in humans is a long way off.

- Cognitive and systems neuroscience investigations also investigate the brain at different levels of abstraction, which correspond to more versus less abstract descriptions (rather than descriptions that are physically microscopic or macroscopic). Figure 4.4 showed this.

- Many studies mix levels and terminology. For instance, one common mistake is to make an unjustified inference equating a neurobiological measure and a psychological concept (for example, amygdala activation equates to feeling fear), and then to use that false equivalence as part of an explanation.

CHAPTER 5

The Neurobiology of Emotion in Animals

GENERAL CONSIDERATIONS

In order to study the neurobiology of emotion in animals, we have to agree that animals have "emotions," according to our usage of the term. As we noted in earlier chapters, some researchers, such as the neuroscientist Joseph LeDoux, have argued recently that the term *emotion* should be used exclusively in its colloquial sense to refer to subjective phenomena ("feelings"), which can only be studied in humans (LeDoux 2012). Animals, he has argued, are useful for studying other internal states that are related to emotions, such as "arousal" or "motivation" (more about those later), but not for studying "emotions" as he now uses the term.

But if we restrict the scientific usage of the word *emotion* to its colloquial (nonscientific) usage, then we have to consider the impact of that semantic choice on the understanding and investigation of this subject among professional scientists (both specialists and nonspecialists), students and the general public as well. "Emotion" is often popularly viewed as distinct from either "robotic" processes—preprogrammed reflexes—or "cognitive" processes—thinking, analyzing, decision-making, and such. While restricting the scientific usage of the term *emotion* to the study of humans does not logically imply that animals lack emotions, it may nevertheless lead many to that conclusion—and thereby create the impression that animals are just robots. That impression would be a false one, and could undermine progress by reducing government research funding for studies of emotions in animals (Fanselow and Pennington 2018). This would greatly limit our ability to understand emotion and apply the benefits of this understanding to the treatment of psychiatric and psychological disorders—not only in humans, but in animals as well. As described further on, modern neuroscience has developed an arsenal of powerful experimental approaches to study practically every type of mental function, including sensation, perception, decision-making, spatial navigation, learning, memory, attention and motor coordination. Importantly, these experimental approaches, such as optogenetics and calcium imaging, (highlighted in boxes 5.1–5.4) cannot be applied in humans for both ethical and practical reasons (box 5.5). Why should emotion be so singular among brain functions that its understanding should be vitiated by making it off limits to experimental approaches that are fundamental to solving almost all other problems in neuroscience?

As we already noted earlier, scientists often use language that has a technical definition that is different from its colloquial usage. We (Anderson and Adolphs 2014), like Darwin (Darwin 1872/1965) and others, therefore use the term *emotion* to refer to a class of *internal brain states*, which are *expressed* by animals and people in a number of measurable ways, including facial expressions, behavior, physiological changes, and (in humans and possibly in some animals) subjective feelings (see figure 2.2). By defining *emotions* as internal brain states, we can study the neurobiological implementation of those states in animals, without concerning ourselves with the question of whether animals do or do not have subjective feelings, which we can continue to study in humans (Dolan 2002; Damasio and Carvalho 2013). We already emphasized this scientific usage of the word *emotion* in chapter 2, but we reiterate the issue here, since it is absolutely critical to keep in mind as you read through the rest of this book.

Scientists usually restrict their study of natural phenomena to things they can measure, but such measures can be more or less direct. At present, we do not have an "emotion meter": a way of measuring emotion states directly in humans or animals in the way that we can measure temperature with a thermometer. Instead, we must infer their existence from measurements of behavioral or physiological variables that express such states, the responses of these variables to experimental perturbations, and a number of background assumptions.

That is not at all unusual in science; particle physicists infer the existence of subatomic particles and forces by measuring the consequences of atom-smashing experiments, using a very complicated series of inferences. Likewise, emotion scientists infer the presence of a fear state by measuring coordinated changes in autonomic variables (heart rate, blood pressure), endocrine variables (levels of stress hormones in the blood), and behavioral variables (freezing or flight, defecation, and so on), in response to a particular situation, and then try to derive the most parsimonious explanation from these, such as inferring a state of fear. Of course, none of these variables measured in isolation is necessarily indicative of an underlying emotion state. But because these and other measurable variables co-occur, in a coordinated manner, with fear in humans, we feel comfortable inferring the existence of a similar internal emotion state when they co-occur in an animal, especially one with a nervous system organized similarly to ours (Davis 1992). These traditional criteria become more difficult to apply to phylogenetically distant organisms, such as insects, that have different behaviors, physiology, and brain anatomy than we do. For that reason, we have endeavored to identify more general features of the expression of emotion states (chapter 3) that can be used to identify and study instances of emotions, or "emotion primitives," in organisms in which homologies to human brain structure and behavior are less obvious.

As we emphasized above and in chapter 2, we believe it is critically important to distinguish between emotion states, and conscious experiences of those emotions ("feelings"). We can study feelings in humans, but at present we do not have any agreed upon criteria with which to investigate feelings in animals; consequently, we will not discuss animal feelings in this book. These considerations are not intended to dismiss the question of subjective feelings in animals. That question is an important and fraught one, because of its implications for the "uniqueness" of humans, and because of ethical considerations. In the future, it may be possible to scientifically address this question; indeed, others (most prominently the late neurobiologist Jaak Panksepp) have argued that current evidence supports the idea that states evoked by electrical brain stimulation in animals reflect emotional feelings (Panksepp 2011b). While the debate over subjective feelings in animals seems unlikely to be resolved soon, there is nevertheless a great deal to be learned about how internal emotion states are implemented in brains by studying animals. In the same way, over the last fifty years neuroscientists have learned a great deal about vision from animal studies, without trying to explain our subjective experience (the "qualia") of seeing the color red, for example. From this perspective, our approach to the neurobiology of internal emotion states is about fifty years behind our investigations of vision.

There is a final important point to make in relation to the question of whether animals have a conscious experience of emotion, one to which we will return in later chapters. This is that it may be possible someday to obtain convincing evidence of subjective experiences from data other than verbal reports. The neuroscientist Adrian Owen, for example, has been able to obtain strong evidence that human patients who were thought to be in a coma have conscious experiences—from imaging their brains with fMRI and comparing the activation to that of healthy people (Owen et al. 2006). Of course, this makes some assumptions, such as that people's brains are fairly similar to one another, and that fMRI is a sufficiently fine-grained method. But still: why should verbal reports be any more reliable indicators of conscious experiences than brain imaging data, once we know enough about how the brain works? If we found markers of brain activity that were very reliably correlated with conscious experiences of emotion, we could certainly use these as evidence for conscious experiences of emotion—even in the absence of any kind of mechanistic explanation. We might still not understand why that particular pattern of brain activation accompanies conscious experiences of emotion, but if the correlation is reliable, the former can be used to infer the latter. For that matter, we do not understand why particular verbal reports accompany conscious experiences of emotion either! This possibility should be kept in mind as we describe the neurobiology of emotion in animals: although animals will never be able to speak to us, it is at least conceivable that we will have brain signatures of emotion experiences. In arguing for a programmatic separation between the study of emotion states and the study of conscious experiences of emotions (as we did at the end of chapter 2), we do not want to imply that the study of animal experiences should be forever off limits. We just need to find the right dependent measures.

Why Do We Need Studies of the Neurobiology of Emotion in Animals?

Given the difficulties in studying conscious experiences of emotions in animals, you might ask "why bother with animal studies at all?" Why not restrict our study of emotions to humans, where there is no disagreement about whether we, as a species, experience subjective feelings? As we will see in chapters 8 and 9, there is certainly a great deal to be learned about the neuroscience of emotion in humans by combining psychological and neuroimaging studies in a rigorously quantitative manner. In some cases, we can actually predict emotional experience from measured patterns of brain activity. If we can make such predictions by looking in the brain, doesn't that say we understand emotion well enough just by investigating the human brain?

Unfortunately not, and the reasons depend in large part on the distinction between correlation and causation, a distinction we clarified in chapter 4. To return to a simple analogy used earlier in this book, we can predict the speed of an automobile by measuring the reading on its speedometer. But that doesn't mean that the speedometer is the *cause* of the car's movement. Indeed, it is not any part of the mechanism that makes a car move; if you take out the speedometer, the speed of the car doesn't change one bit. But the speedometer has strong predictive value, nonetheless. To understand emotion, we need to move beyond prediction and correlation, to causation; we need to discover the engine in the car and figure out how it works. To do that, we need to take it apart, piece by piece, in order to understand the function of each of its components and explain how they work together. It is not currently possible to do that for conscious experiences in animals, and it is very difficult to do that for *any* aspect of emotion in humans, but it can be done for the investigation of emotion states (not their conscious experience) in animal studies.

This process of mechanistic (causal) understanding of brain function requires *perturbation*, in addition to careful, quantitative observation.

Perturbation means going into the brain (performing *invasive* experiments) to deliberately and specifically manipulate the activity of its genes, neurons, or circuits, and to measure the consequences of those manipulations. In humans, such invasive experiments generally cannot be performed systematically, with the necessary precision and reproducibility; they are restricted to uncontrolled rare accidents (such as the case of Phineas Gage, which we discuss in chapter 8), to procedures rigidly constrained by medical ethics for a curative or therapeutic justification (such as stimulation through depth electrodes), or to noninvasive methods that are far too coarse (such as transcranial magnetic stimulation).

A more specific reason that we need animal studies of emotion is that we can apply powerful new techniques, such as optogenetics or pharmacogenetics (box 5.1) and optical imaging of neuronal activity (box 5.2), in a manner that is targeted to genetically defined neuronal cell types (box 5.3). These recently developed methods, which afford a far greater level of cellular and temporal resolution than do fMRI, lesions, or electrical stimulation, cannot (yet) be applied to humans (see box 5.4); moreover, even if they could, their application would be restricted by medical necessity. The tools that can be applied in animals, which are detailed in the boxes for ease of reference, highlight an important level of precision and discrimination that was difficult to convey clearly in our earlier figure 4.2.

Central to these methodologies is the concept of neuronal cell types (box 5.3). Neurons do not come in a single "flavor"; there are many different types, or classes (Zeng and Sanes 2017). The simplest distinction is between excitatory neurons and inhibitory neurons. However, within each of these categories there are multiple subtypes; for example, a recent study showed that primary visual cortex (area V1) in mice contains at least twenty-five different subtypes of excitatory neurons and an equal number of inhibitory neuron subtypes; the total number of cell types in this cortical region alone may exceed one hundred. Neurons of a common type typically express similar sets of genes and hence they have similar profiles of receptors and ion channels; they also have similar patterns of connectivity with other cell types. It is therefore likely that these neuronal subpopulations are a functional unit of sorts: neurons of a specific subtype all "do the same thing." This is why manipulation of such a defined population of neurons in a given region, such as the amygdala, can have one type of effect on an emotion (as we will describe in chapter 6), while manipulating a different population of neurons in the same brain region can have the opposite-direction effect. Different neuronal types may thus constitute the smallest meaningful functional unit of cellular organization in brains and may show considerable conservation across the brains of different species. We do not yet know the total number of different cell types in the mammalian brain; current estimates run in the thousands (even the retina contains ~60 different cell types). But that's still a lot less than the total of 80 billion or so neurons in the human brain. In a way, this is good news for a neuroscience of emotion.

Importantly, different neuron types are often intermingled within a given brain region, like salt and pepper—they are not necessarily all packed together in a specific, easily identifiable structure or nucleus (note: the term *nucleus* is used here in the neuroanatomical sense to refer to a brain structure, not in the cell biological sense to refer to the organelle that contains the chromosomes). In order to accurately measure patterns of neuronal activity in the brain, it is necessary to assign the activity to specific cell types, which requires the tools we discuss in boxes 5.1–5.3. So, even if fMRI could resolve individual neurons (which it can't), it could not tell the difference between them; it is a low-resolution *and* a "monochromatic" method. (Also monochromatic are conventional c-fos analysis, or extracellular electrode recordings, which *do* have single-cell resolution.) By contrast, using genetics to express calcium sensors selectively in a specific class of neuron makes it possible to visualize neuronal activity with single-cell resolution, *and* to selectively visualize the activity of a specific *type* (or types) of neuron (box 5.2). As we shall see in chapter 6, the application of these techniques in mice has led to a replacement of the classical view of neural networks as connections between anatomical structures (for example, between "the amygdala" and "the prefrontal cortex"), with one based on connections between different cell types, where a given structure may contain multiple cell types with different anatomical connections.

Ultimately, the level of mechanistic understanding afforded by animal experiments will be crucial for developing new therapies for psychiatric disorders—a hugely important task given the enormous burden that such disorders impose on society, and the fact that the pharmaceutical industry has largely abandoned the search for new psychiatric drugs (Griebel and Holsboer 2012). From our perspective, the fact that animal studies of emotion have not led to a new psychiatric drug in the last half century is not, as some have argued (LeDoux 2015), because animals are poor subjects for testing drugs that influence emotion states; rather, it is because emotion is an enormously complicated brain function, and we have not previously had the tools to study it with the necessary level of sophistication and depth. By way of example: the new anticancer drugs that seem to be so promising reflect decades of research on the detailed underlying mechanisms of cell division, cell death, and immunologic function in animals, despite the fact that animals have different physiology and immune responses than humans. Why should understanding emotion be any different? We return to this point at the end of this chapter.

BOX 5.1. Optogenetics and pharmacogenetics.

These terms refer to methods developed to manipulate the activity of genetically marked subclasses of neurons within a given brain region (see box 5.3). They are most commonly used in organisms where germline transgenesis is feasible, including the nematode worm *C. elegans*, the vinegar fly *Drosophila*, zebrafish, and mice. Their application to other organisms such as rats, nonhuman primates, and humans is currently more limited (although not impossible). In these methods, a gene encoding an engineered microbial light-sensitive ion channel or pump ("opsin," or optogenetic effector), or an engineered drug receptor (pharmacogenetic effector), is expressed in a desired subclass of neurons, using viral vectors or a heritable transgene. Depending on the type of effector used, the neurons can then be activated or inhibited at will in the freely moving animal.

Optogenetic effectors are actuated using light of a particular wavelength, delivered into the brain region of interest via an implanted optic fiber. Pharmacogenetic effectors, called DREADDS (see glossary), are actuated by administration of a drug that binds to an engineered receptor; this drug does not activate any endogenous receptors, nor do any endogenous ligands activate the synthetic receptor. Both methods allow functional manipulations to be targeted to a specific population of neurons, even if those neurons are intermingled in a salt-and-pepper-like manner with other classes of neurons within a given brain region. This specificity is crucial in brain regions like the amygdala central nucleus that contain mixed populations of neurons exerting opposite-direction effects on the same behavior: if both populations are simultaneously activated or inhibited, e.g., using conventional drugs or electrical stimulation, the manipulations may cancel each other out and no effects on behavior may be observed. Importantly, unlike conventional lesions, these methods are rapidly reversible, albeit on different timescales: optogenetics allows activation or inhibition of neuronal activity to be performed using laser light pulses with millisecond resolution, while pharmacogenetics allows reversible manipulations on a timescale of tens of minutes or hours. These approaches can also be applied together, e.g., activating one population of neurons optogenetically, while pharmacogenetically inhibiting a putative target of those neurons.

The application of these methods has had a transformative influence on the ability to infer causal relationships between the activity of specific neurons and behaviors, by testing their sufficiency (using activating manipulations) or necessity (using inhibitory manipulations; see chapter 4). Nevertheless, like any method, these tools need to be applied judiciously, and the results interpreted with caution. As mentioned in chapter 4, the demonstration that optogenetic activation of a given brain structure or cell population is "sufficient" to evoke a particular behavior does not imply that it does so *directly*; there may be many unknown intervening circuit elements. Furthermore, optogenetic stimulation can impose highly artificial patterns of synchronous activity on a population of neurons, which the cells would never exhibit under normal circumstances; this could potentially yield artifactual effects on behavior. New methods are being developed to optogenetically activate neurons in a more spatiotemporally controlled manner, designed to mimic their normal pattern of activity. This approach requires integrating optogenetics with methods to measure the activity of the same neurons, e.g., using calcium imaging (box 5.2). Optogenetics and pharmacogenetics also offer the potential to develop new therapies for psychiatric disorders based on targeted manipulation of specific neuronal populations or circuits; in essence, a more specific form of deep brain stimulation (DBS). However, the application of these tools in humans faces significant regulatory hurdles, in part because they require expressing foreign genes in the brain (which may itself promote unwanted side effects).

BOX 5.2. Optical imaging of neuronal activity with genetically encoded reporters.

Measuring neuronal activity in vivo at single-cell resolution has traditionally been performed using electrophysiological recordings. While the measurement of action potentials ("spikes" of electrical potential through which neurons communicate with one another) remains the gold standard for neuronal activity measurements, it has a number of shortcomings: (1) It is typically measured, in intact freely moving animals, using extracellular electrodes. Therefore, there is no easy way to fill and identify the cell(s) after recording; furthermore, subthreshold changes in membrane potential cannot be measured. (These problems can be overcome using intracellular or patch-clamp recording, but that is much more difficult to do in freely moving animals.) (2) It is difficult to sample the same cell across multiple days of recording, due to instability in electrode placement. (3) Recorded neurons have to be sampled at relatively low density in order to maintain electrical isolation between recording sites; therefore, measuring the activity of high-density ensembles is challenging. (4) It is challenging (but not impossible) to assign electrophysiological measurements to specific, genetically identified neurons.

These limitations can be overcome by using genetically encoded calcium and voltage indicators (GECIs and GEVIs, respectively) to optically measure neuronal activity. These sensors are protein molecules (originally isolated from jellyfish or microorganisms) that have been engineered to report changes in intracellular free calcium (which increases with neuronal activity), such as GCaMP6, or changes in membrane voltage, such as "ArcLight," as changes in fluorescent emissions. They can be genetically addressed to specific populations of neurons using the same type of methods employed to express optogenetic or pharmacogenetic effectors (box 5.3). Therefore, they allow the activity of genetically defined neuronal subpopulations or cell types to be directly measured. Furthermore, because expression of the indicators is stable, the same cells can be imaged across multiple days, provided that the same microscopic imaging field can be identified. In addition, high-density populations of neurons can be recorded using these optical methods, provided sufficient microscopic resolution. Finally, while GECIs are best suited to the detection of calcium influx accompanying action potentials, GEVIs in principle allow detection of subthreshold events, given adequate sensitivity. These methods provide an important complement to (but are not intended to replace) electrophysiological methods. The sensitivity and temporal resolution of GECIs and GEVIs continues to be improved. Different color GECIs and GEVIs are being developed, which should allow simultaneous imaging of genetically distinct neuronal subsets within the same brain region. A key aspect/limitation of optical imaging is the imaging methodology itself. This ranges from bulk fluorescence measurements of population activity using optic fibers in freely behaving animals (so-called fiber photometry), to single-cell resolution imaging in head-fixed preparations using two-photon microscopy. An important new advance is the development of miniature head-mounted microscopes coupled to long, thin (0.5–1 mm diameter) glass microlenses, allowing imaging in deep brain structures such as the amygdala or hypothalamus, in freely moving animals. Other methods, such as light sheet or lattice light sheet microscopy, allow for the simultaneous imaging of tens or hundreds of thousands of neurons in a single, transparent specimen (such as a zebrafish larva), but those must be head-fixed beneath the microscope lens. Combined activation or inhibition with imaging of neuronal activity is feasible, but spectral overlap between optogenetic effectors and GECIs or GEVIs remains problematic.

BOX 5.3. Genetically defined neuronal cell types.

The brain contains many different classes, or "types," of neurons. Many of these cell types can be easily distinguished by their morphologies (how they look under a microscope), such as Purkinje cells versus granule cells in the cerebellum. However, they can also be distinguished by differences in the profile of genes they express: a typical (mammalian) neuron may express 10 to 15,000 of the total ~25,000 genes in the human genome; neuron types differ in which of those 10 to 15,000 genes they express. An advantage of using genes to identify different cell types is that genes can be used, in tractable model organisms, to gain access to different cell types in order to mark, map, and functionally manipulate them. This is possible because the expression pattern of any given gene is determined by sequences of noncoding regulatory DNA called enhancers or silencers (these elements are often referred to, incorrectly, as "promoters"; this latter term correctly denotes the site that determines the start of mRNA transcription for a gene). If the enhancers that control the expression of a specific gene are physically linked to the DNA sequence of a genetically encoded neuronal indicator or effector (GENIE), and if those linked sequences are inserted into the genome of an organism (or just into the genomes of neurons in its brain), then ideally the GENIE will be expressed only in those cells that express the specific gene. For example, a subset of inhibitory neurons in the mammalian cortex expresses the calcium binding protein parvalbumin (PV);

PV is therefore considered a "marker" for that class of neurons. If a transgenic mouse is created in which regulatory sequences from the PV gene are fused to the gene encoding GCaMP6, then in the brain of that mouse only PV$^+$ neurons will express GCaMP6, allowing calcium imaging of just that population. This approach can be generalized to express any kind of GENIE (or any other foreign gene) in the neurons of interest. However, the process of identifying those regulatory sequences can be difficult, laborious, and time consuming. Therefore, many researchers have opted to take a complementary approach in which the DNA encoding the GENIE of interest is simply inserted into the chromosomal sequence of the marker gene, using a method called *homologous recombination*. This allows the GENIE to be expressed in exactly the same way as the marker gene itself. However, this method requires introduction of the GENIE sequence into the germline of the organism, a process that is technically difficult in many organisms and currently prohibited in humans for ethical reasons. Alternative methods for expressing GENIEs, such as using disabled viruses as vectors ("disposable molecular syringes" for delivering DNA to cells), are possible, but those present their own challenges. A popular approach is currently to use a "binary system" in which a line of transgenic mice is made that expresses an activator gene (e.g., "Cre recombinase") in the cell population of interest; this mouse line can then be injected in its brain with viral vectors containing a particular GENIE that is expressed only in response to the activator gene. In this way, the same mouse "Cre line" can be used to express dozens of different GENIEs in the cell population of interest. However, in many cases a single marker gene may not be sufficient to distinguish one type of neuron from others in a given brain region. In that case, more cumbersome "intersectional" genetic strategies, based on combinations of marker genes, are required to restrict GENIE expression to the cell type of interest. Although we are currently years away from having a complete catalog of the different genetically defined cell types in the mammalian brain (which likely number in the thousands), this process is being speeded up exponentially through the application of single-cell RNA sequencing methodologies (scRNA-seq) to neurons (Zeng and Sanes 2017).

What Do We Want to Understand about Emotion by Studying Animals?

Our objective, in animal studies as in human studies, is to understand how emotion states are produced by causal mechanisms in the brain. At one "nuts-and-bolts" level, we would like to provide an account (a narrative) of what happens in the brain when an animal is exposed to an emotional stimulus, in terms of the cell types, circuits, neuromodulators, and synaptic connections that are engaged. We want this description to incorporate causality (necessity and sufficiency) as measured by perturbation experiments as well as to provide a detailed description of the patterns and dynamics of neural and chemical activity as measured by electrical or optical recording. We want it to explain how the different "read-outs" of a given emotion state are generated: what causal chain explains the production of emotional facial expressions, autonomic responses, and emotional behaviors?

We also want our explanation to constrain computational and functional models of the emotion state, which typically happens as we test our actual data against the predictions made by those models. We want to understand how the different properties of emotion states, such as valence, scalability, persistence, coordination, and generalizability (see chapter 3 and figure 3.2) come about: how are they implemented in neural circuits? Ultimately, we want our account of the causal neural mechanisms that instantiate an emotion state to be so complete and transparent that it lets us understand how all the higher-order features of the emotion are produced, and it lets us understand how these achieve the biological function of the emotion for that animal.

To organize our understanding, we need to be able to relate different findings to one another. We would like to understand the "similarity space" from which different emotion states are constructed: what are the dimensions of this space, and how are they represented in the brain? Are there anatomically distinct brain structures that coordinate or organize each type of emotion state, and if so, how many are there? Or is each emotion state assembled from a more distributed set of circuits that partially overlap?

We revisit here the issue of a similarity space, since it is critically important (see also the discussions in chapter 3 and chapter 8). This concept can be appreciated by analogy to color representation in the brain. We know that the "similarity space" that defines all the colors our brains can perceive has three dimensions, and that they are red, green, and blue (RGB). We understand that the neurobiological basis of these dimensions is based on the spectral sensitivity of three different classes of photoreceptors in the retina, which is in turn irreducibly defined by the molecular properties of the specific rhodopsin protein that each class of photoreceptor expresses. (Note that it did not have to be this way; colors could also have been represented by a four-dimensional space, for example, CMYK like some printers use, but evolution did not produce this solution.) In the case of color, the similarity space is simple in retrospect, but that is not always the case for primary sensory systems. For example, there is ongoing debate about the dimensionality of the similarity space that allows our brains to discriminate different odors.

Fine, you may say, but why can't we just study the neural instantiation of each emotion state on its own, without getting into the issue of similarity space and dimensionality? Indeed, that is how much emotion research in animals has proceeded so far (human studies have actually investigated similarity spaces—but only with respect to subjective judgments of the feelings of the participants, not objective criteria of the scientist doing the research). But consider how this approach would limit our understanding of color perception if we did not know that different colors are represented by linear combinations of RGB in the retina. For example, could we understand how the color "orange" is encoded? We could begin by looking for orange-selective cells somewhere in the visual cortex, and indeed we might find some. If we found such cells, we would be very excited, but then we would immediately want to know whether there are cells selective for other colors—magenta, violet, or puce, for example. But how many such cells should we be looking for? There is an almost infinite gradation of different colors across the spectrum; color is a continuously varying property. Since there aren't an infinite number of cells in the brain, we would start systematically measuring the "tuning curves" of the color-selective cells and compare them, analogous to measuring the orientation tuning of neurons in V1. Once we established such color-selective tuning curves, we would soon realize that the brain's ability to distinguish a large variety of colors must ultimately be based on linear combinations of a more basic set of different color-selective cells.

So yes, we could start by investigating how a given color is represented centrally in the brain, and we might well find evidence for a representation of that particular color, such as orange. But it would be difficult to move beyond that phenomenology without considering how *different* colors are encoded—which brings us back to the issue of the "similarity space" or dimensionality of color representation. In the same way, it seems difficult to escape the need to understand how *different* emotions are represented in the brain, even if we just want to understand how any one emotion is represented or encoded.

Of course, there are historically famous (and also contemporary) examples where specific emotional behaviors can be elicited by stimulating specific regions of the brain (see Panksepp et al. 2011b)—most notably Walter Hess's Nobel Prize–winning demonstration that "defensive rage" can be evoked in a cat by stimulation of the lateral hypothalamus (Hess and Brügger 1943). But, viewed in the context of the color analogy, that is equivalent to discovering a cell in the visual cortex that, when stimulated, makes you see orange. That would be an amazing result, and an important first step. But ultimately, it would be difficult to move beyond that observation without considering how the brain encodes, represents, and discriminates different colors. The same would seem to apply to different emotion states. We do not want merely to be "stamp collecting" by finding neurobiological substrates of individual example emotions; we want to generate a predictive understanding of emotions in general. Such an understanding would ultimately allow us to read out a specific type of emotion state from the pattern of brain responses, and, conversely, to predict what the brain response would be for an emotion state evoked by a particular stimulus. Such an understanding might also ultimately allow us to build a robot that has emotions, at least in principle.

The Relationship of Emotion States to Motivation, Arousal, and Drive

Darwin believed, and we agree, that emotional expression (whether produced in the face, the body, or both) was an evolutionarily conserved function, and that its particular manifestations in different species provide insights to how emotions evolved (Darwin 1872/1965). He attempted to explain these expressions in terms of the adaptive value that particular expressions afforded humans and animals. An important and outstanding question, which Darwin did not address, is how emotions evolved and when in evolution they first appeared.

There are several parts to this question. First, when did different types or specific emotions evolve; for example, did fear evolve before shame (a social emotion)? Do you need to have some emotions before you can evolve others? Is there some kind of logical order in which all the different varieties of emotions evolved in phylogeny?

Second, when and how did emotions begin to evolve from simple stimulus-response reflexes? What kind of intermediates, "emotion primitives," might we see in the progressive evolution from reflexes to emotions proper? Are emotion-like internal states, as we have defined them, a property of all multicellular organisms with a nervous system— including jellyfish? Or can we define key transitions in the evolution of emotions from reflexes?

Third, is a given emotion, such as "fear," represented in a similar manner in phylogenetically distant nervous systems? Are there general principles of neural circuit architecture that underlie emotion states, at a sufficient level of abstraction, even if the details of their instantiation will no doubt vary between distantly related species? For instance, would "fear" roughly fall within the same region of a dimensional space defined by something like the list of figure 3.2 in all species? These are very much open questions, and answering them will require us to survey emotion states broadly, with a consistent terminology.

Such a comparative investigation will also help us to delimit emotion states from other central states that seem similar to emotions. Some of the key emotion properties such as valence, scalability, persistence, and generalizability surely apply to other internal states as well, such as those referred to by psychologists and ethologists as "motivation," "arousal," or "drive." There is general agreement, even among those who differ on the definition of "emotion," that motivation, arousal, and drive ("MAD") states can be studied in animals (Berridge 2004; LeDoux 2012). So, if we apply our general emotion properties as criteria (chapter 3) to try to identify instances of emotional expression across phylogeny, how do we know that we are studying emotion states (as we have operationally defined them) and not MAD states?

At one level, this is just a semantic issue—we don't see a problem if there are also other states whose properties may overlap with those of emotions. If we are interested in the evolutionary origins of emotion states, and seek to understand how emotion primitives are instantiated in different kinds of brains, then it does not matter what we call these states, so long as we are specific about their properties (persistence, scalability, valence, and so on) and have accurate and objective ways to measure them. Indeed, one of us has introduced the neutral term "π states" to refer to internal animal brain states that exhibit persistence and scalability, so as to avoid unproductive semantic arguments about whether these states need to be called emotions, motivations, or drives (Anderson 2016). Moreover, to the extent that arousal and perhaps motivation are themselves dimensions or features of emotions, studying the neurobiological mechanisms that underlie such features is very relevant to understanding emotion. By way of analogy: one of our Caltech colleagues recently redefined Pluto as a "planetoid" rather than a planet (Brown 2010). Did astronomy change as a consequence? Not really—the redefinition was an interesting taxonomic update that grouped Pluto with other more recently discovered planetoids in our solar system, but it didn't change our scientific understanding of Pluto.

All that said, many theorists care very much about the difference between "real emotions" and MAD states, especially those for whom the essence of emotion is its subjective, conscious nature (LeDoux 2012; Feldman Barrett 2015) (figure 5.1). (We note that such an essential feature could also apply to motivations or drive states, since humans clearly have a subjective feeling for states such as "hunger.") Therefore, it is important to understand the differences between these states as operationally and conceptually defined by others. As these definitions

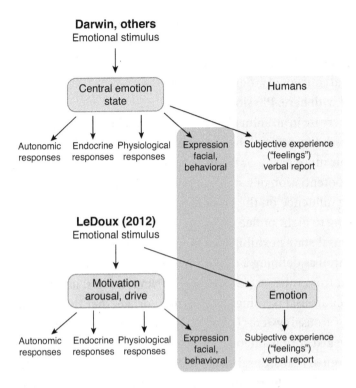

FIGURE 5.1. Different ways of thinking about emotions and MAD states. Darwin viewed facial expressions, body posture, and observable behavior (pink rectangle) as "expressions" of emotion states that are conserved between humans and animals. Other writers (e.g., LeDoux 2012) have argued that these observables are *not* expressions of "emotions"—according to their definition—but rather are expressions of related but distinct internal states such as motivation, arousal, or drive, which can be measured in animals as well as in humans. In their view, subjective feelings, which can only be accessed by verbal report in humans (gray rectangle), are the only acceptable evidence of an "emotion"; ergo, emotions are only attributable to humans. In Darwin's (and our) view, verbal reports are but one of many different "read-outs" (measurable expressions) of central emotion states, but are not an essential defining component. Our view does not, however, deny animals subjective feelings; it simply says that evidence of such feelings is not essential to infer an emotion state in an animal.

have been discussed elsewhere in more detail (Berridge 2004; LeDoux 2012), we will only briefly survey MAD states here.

Arousal

As we already noted in chapter 3, psychological theories of emotion typically use *arousal* to refer to a basic dimension describing people's ratings of how they subjectively feel; that dimension reflects the intensity or strength of the subjective experience. That usage of *arousal* is certainly closely related to, but conceptually quite distinct from, a physiological concept of arousal, which is the concept of arousal we are concerned with here. Physiological arousal states are typically manifested as an increase in an animal's muscle tone, and in a lower threshold for its responses to sensory signals. Arousal states can cause certain behaviors to occur more easily or frequently than they otherwise would (such as the potentiation of a startle response under threat) and may exert a biasing influence on the choice between different behaviors (such as choosing to flight or flee during an escalating aggressive encounter).

Arousal states exhibit persistence and scalability but are usually thought of as defining a dimension of affective similarity space that is distinct from valence (see chapter 3). Nevertheless, it remains unclear whether arousal is a single dimension, or whether there are different types of arousal systems (partly orthogonal dimensions of arousal) that are engaged during different behaviors (Lebestky et al. 2009). It has also been argued that arousal states are themselves intrinsically affective in nature, and specific to different primary emotion systems such as rage or fear (Panksepp 2011a). While motivation states are often accompanied by increased arousal, arousal can occur without motivation, for example, the increased arousal that occurs during a sleep-to-wake transition.

"Generalized arousal" in vertebrates is commonly attributed to the release of neuromodulators such as norepinephrine (NE), which is produced in a specific brainstem nucleus, the locus coeruleus (LC). Neuromodulators are small protein-like molecules that function as neurotransmitters of a special sort. Like classical neurotransmitters (molecules like GABA, or glutamate), neuromodulators serve to signal between neurons, but they tend to do so in a more diffuse fashion and typically act in concert with (modulate) the action of classical neurotransmitters. Arousal can also involve other neuromodulators, such as dopamine (DA), acetylcholine (ACh), and other small molecules (Pfaff, Westberg, and Kow 2005). Even sex steroid hormones and neuropeptides can promote arousal states. For example, the neuropeptide hypocretin (also known as orexin) is released from a small group of hypothalamic neurons, whose optogenetic activation can promote sleep-wake transitions and whose impairment can produce narcolepsy, a disabling condition in which people suddenly fall asleep without warning. Therefore, it is likely that arousal is far more multi-dimensional and complex in its neural instantiations than is commonly believed.

How the scalability and persistence of arousal states are controlled is not well understood, either. A common assumption is that the intensity of an arousal state simply scales with the amount of the relevant neuromodulator that is released. But the evidence for that is slim. Furthermore, even if it were true, it is not clear how such a scalable increase of neuromodulator release would be controlled (for example, via more NE neurons activated, or more active NE neurons, or different subpopulations of NE neurons activated, and so on), and whether it would be distributed across the entire brain (again, a common belief based on the broad and diffuse projections of NE neurons), or whether it would target specific brain regions. Recent observations suggest that NE neurons are actually heterogeneous and that different subpopulations of them may project to different brain areas.

Finally, the control of the persistence of arousal states is also not well understood. It may reflect persistent activity of neuromodulatory neurons that promote arousal, the chemical stability of the modulators once they are released, or persistent activity in their downstream targets. Thus, while "arousal" is typically conceived as a rather simple, "hydraulic" sort of brain property, it seems likely to prove far more complicated and will require a lot more research before we have a grip on the underlying mechanisms. Which of the above aspects of physiological arousal correspond best, or have causal relations, to the subjectively experienced dimension of arousal is unclear—you can get people to rate different varieties of "arousal" depending on exactly how you ask the question.

These considerations also raise a theme that pervades all of this book, and that will become apparent again when we dig into the neurobiology in the next chapters. The detailed neural mechanisms that underlie emotion states are very, very complex. This is perhaps not too surprising—the initial functional schemes and psychological categories that we have available are based on much more macroscopic data and have to a large extent drawn on our folk psychological concepts and categories, which are extremely coarse. But given the complexity of the details, there is certainly a strong sense that the neurobiological details may entail fairly radical revision or replacement of the terms and concepts that we currently have available. Put another way, once we have sufficiently detailed data about the neurobiological mechanisms, we may come up with algorithmic models that are considerably more varied and have a considerably larger number of parameters than do our current models. Similarly, once we push these models up to the functional level, the story of what emotions do will turn out to be a lot more nuanced and complex than we could have imagined. We have no doubt that there will be such a story at the functional level, but we may well realize that evolution has selected a larger number of more complex mechanisms. Again, this is probably not a big surprise to biologists, and certainly not to neurobiologists. One valuable point to make is that such a fine-grained functional story might be difficult or impossible to glean simply from ethological and comparative behavioral studies. Such studies are informative in providing an initial scaffolding on which an evolutionary story can be elaborated, but in the absence of a time machine, this story will always be underdetermined. Neuroscience data can help fill in the details, bit by bit. In other words, "bottom-up" as well as "top-down" approaches to the study of emotion at different levels of abstraction (figure 4.4) will be important.

Motivation versus Drive

As with "arousal," "motivation" and "drive" are also scalable internal state variables, with the key difference from arousal being that they promote goal-directed behavior. While states characterized by strong motivation or drive are often accompanied by elevated arousal, they need not be; for example, a fatigued person (low arousal) may nevertheless be highly motivated to go to bed. The difference between "drive" and "motivation" is more of an operational and conceptual one than a biological one.

Ethologists such as Konrad Lorenz conceptualized drive as a constitutive "hydraulic" internal state, whose "pressure" provided an intrinsic

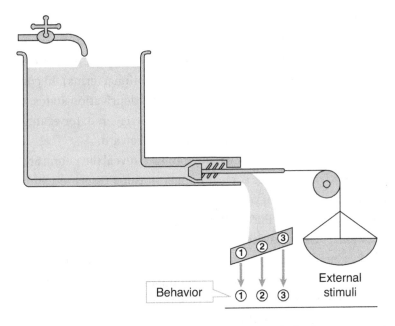

FIGURE 5.2. Konrad Lorenz's hydraulic conception of drive. The hydraulic pressure producing "drive" in this metaphor is assumed to be produced internally and constitutively (faucet), rather than being triggered by the presence of an external stimulus. However, Lorenz did acknowledge the possibility of external triggers in later versions of his theory. "1, 2, and 3" are different behaviors that are "released" at different levels of hydraulic pressure and/or strength of external triggers. Figure modified with permission from Anderson 2016.

impetus to perform a behavior, for example, to mate. In order for the behavior to be expressed, an external signal had to "release" this pressure, allowing it to flow through the nervous system and drive the behavior (figure 5.2). Once the behavior was performed, the hydraulic system was drained and would have to be replenished before the drive to perform the behavior could be reestablished. In this view, drives are an internal "force" that propels behavior, a kind of "nervous energy" that builds up and has to be released. Drives are closely related to motivations, and, like motivation, often serve a homeostatic function that may be triggered by interoceptive signals like hunger or thirst (this homeostatic feature has sometimes been highlighted by theorists in distinguishing these states from emotion states). However, drives are not necessarily created by homeostatic imbalances; for example, the "nesting drive" is not created by the deprivation of opportunities to nest. A key difference is that drive states, as conceptualized by Lorenz and others, are not *reactions* triggered in response to a stimulus (or to deprivation of a resource); rather, they are endogenously generated (through "interoceptive" mechanisms) and create a pressure to act, which is "released" by the stimulus (in operational terms). In contrast, motivations may be internally triggered by deprivation states, via interoceptive cues, or they may be externally triggered, for example, by the presence of an incentive that predicts a reward.

At present, while evidence is beginning to reveal the neural architecture of motivation states such as hunger or thirst (Sternson and Eiselt 2017), the same cannot yet be said for "drive" states (at least as conceptualized by Lorenz). Part of the reason for this is that psychology has operationalized "motivation" as an internal state that can be experimentally measured using incentive-guided instrumental (operant) learning paradigms. In contrast, there are no comparable assays for "drive" states grounded in the methods of experimental psychology.

To some extent, the difference between "drive" and "motivation" may be largely semantic: while psychologists and ethologists may argue about what they mean, the brain may not "know" the difference. Put another way, although "motivation," "arousal," "drive," and "emotion" may be useful as heuristic classification schemes to conceptualize and differentiate various kinds of internal states, there is no fundamental law of nature that says that these states must exist as distinct and dissociable patterns of brain activity, or that all internal states related to behavior must fall into one of these categories. This is why we take a relaxed attitude to the possible distinctions between emotion states and MAD states: whether or not you believe they need to be distinct really makes no difference to what we write here about the fundamental properties of emotions.

Motivation versus Emotion

The difference between motivation and emotion is a fairly subtle one, and in part a semantic issue. For example, is "hunger" an emotion? Most psychologists would say no; hunger is a "motivational" state or a "homeostatic" state. However, it has a number of features in common with emotions (including, in humans, a subjective feeling of the state), to the extent that at least one investigator has referred to hunger, thirst, pain, and itch as "homeostatic" emotions (Craig 2002): emotions that function to reestablish stable "set-points" that define optimal physiological conditions such as core body temperature, metabolic energy, and water content. They do this by promoting goal-oriented behaviors that restore the imbalances in the system; for example, hunger promotes foraging (finding food) and consummatory behavior (eating food) in order to restore energy deficits caused by caloric deprivation (that is, starvation). Motivation can also promote behavior to restore imbalances caused by surfeits, rather than deficits, in physiological parameters: hot weather causes animals to seek shade (or humans to seek air conditioning), in order to lower core body temperature and prevent dehydration.

However, as Darwin noted (Darwin 1872/1965), emotions also promote goal-oriented behavior: fear promotes freezing or flight to achieve the "goal" of escaping predation. Nevertheless, emotions may be more flexible than motivations in the range of the behaviors they can promote, depending on context; by contrast, motivations tend to be more specific in the behaviors they promote (for example, "find and consume food or water"). Moreover, in contrast to emotions, which are typically triggered by exteroceptive (external) sensory cues (for example, the presence of a predator), motivations are usually triggered by interoceptive (internal) cues that signal a homeostatic imbalance in the body (although motivation can also be triggered by external stimuli such as money). Another important difference is that a motivational state (like a drive) is typically terminated after the goal object is attained: we no longer feel hungry after we have eaten. In contrast, emotion states can persist after the "goal" is attained (we may continue to feel residual fear for an extended period of time even after escaping a dangerous encounter with a predator, or feel persistent anger following a violent verbal or physical altercation).

Despite these differences, motivational states exhibit several of the general features we have outlined as properties of emotion states in chapter 3. These include valence (a motivation state can be experienced as pleasant or unpleasant); scalability (the intensity of a motivation state may be lower or higher, depending on the level of "need" or "want"); persistence (motivation states can be long-lasting; however, as mentioned, they generally are rapidly quenched after homeostasis is restored); priority over behavioral control (if you are hungry you will stop playing chess and go to the refrigerator, and the hungrier you are the harder it becomes to counteract this motivation); and multi-component effects and their coordination (for example, hunger causes physiological and hormonal as well as behavioral changes). However, in contrast to emotions, motivations usually do not serve a social communicative function (except in the case where a hungry infant cries to attract its mother to feed it). Furthermore, their participation in learning is less direct than that of emotions: while hunger can probably enhance learning, for example, by improving the association between a physical location and the presence of food, it is not yet clear whether the hunger state itself (need for food) can be conditionally evoked by an otherwise neutral stimulus in a satiated animal, in the same way that a conditioned stimulus can evoke a state of fear (Fanselow 1984). Nevertheless increased feeding can be promoted in satiated animals by exposure to a CS previously associated with food.

When we understand the brain more completely than we presently do, it may prove to be the case that psychological constructs like "motivation," "arousal," "drive," and "emotion" do not map onto cleanly separable neurobiological mechanisms. Indeed, at some points these words may no longer be necessary to explain how the brain allows an organism to adapt to its environment by controlling its behavior, once we have understood these brain functions in sufficient mechanistic detail. By analogy, terms like "specification," "determination," and "commitment" were developed by Nobel laureate Hans Spemann to explain how primitive embryonic cells become progressively locked in to a particular developmental fate, based on his classic embryo transplantation experiments performed in the 1920s. But these concepts were neither subsumed by, nor were they necessary for, understanding the molecular and cellular mechanisms of embryonic development that were laid bare by later genetic analysis of pattern formation in *Drosophila* and *C. elegans* in the 1980s (Lawrence 1992). Terminologies may come and go, but the underlying biological, causal mechanisms that we wish to understand remain the same; the important question is how best to assess when we have understood that which we have sought to explain.

Psychiatric Drugs, Animal Models, and Emotions

One way to assess whether we have correctly understood a biological process or mechanism is to see whether we can use that information to develop a new treatment for a disorder involving that process in humans. In this respect, using animal models to develop drugs to treat psychiatric disorders, such as depression, has been frustratingly unsuccessful. Most antidepressant drugs, including selective serotonin reuptake inhibitors (SSRIs) such as Prozac, were discovered by accident. Although they can be effective, because they are relatively nonspecific they can also promote unpleasant side effects (for example, loss of normal sexual function or dampening of creativity) that lead many people to discontinue their use. The brilliant novelist David Foster Wallace, author of *Infinite Jest*, discontinued the medication he was taking to combat depression because of its side effects and suffered a relapse that led to his suicide at age forty-six.

Despite the inadequacies of current therapies, there has not been a fundamentally new medication for a psychiatric disorder in the last fifty years; indeed, most pharmaceutical companies have abandoned their psychiatric drug-development programs, because of costly clinical failures. Why is this the case? One explanation, as argued recently by Joe LeDoux and Danny Pine (2016), is that the rodent behavioral tests traditionally used to assess the efficacy of psychiatric drugs are misleading, because these drugs are intended to treat disorders of subjective feelings in humans ("I feel overly anxious, or depressed"), and such feelings cannot be measured directly in animals. Although some antianxiety (anxiolytic) drugs may have effects in animals (figure 5.3B, ①), according to this view they are not necessarily affecting the same brain processes as give rise to subjective feelings of anxiety in humans (figure 5.3B, ②).

While this argument may be valid, it violates Occam's razor, a fundamental principle in science sometimes referred to as the principle of parsimony. According to this principle, if there are two competing explanations or theories for an observed process, the simplest one is usually correct. Thus, a simpler explanation for the effects of anxiolytic drugs in humans and animals is that they share a common, core mechanism of action in both species (figure 5.3A, ①), rather than two different mechanisms (figure 5.3, ① and ②). This would explain why some drugs that reduce anxiety in humans can calm down seemingly anxious pets. However, this simpler model (figure 5.3A) implies that animals can share with humans a common central state of anxiety, which is subjectively experienced in humans (and may or may not be in animals; figure 5.3A); in other words, that humans and animals have similar emotion states. But this conflicts with the view that animals do not have emotions (figure 5.1), argued by the ethologist Niko Tinbergen to be more parsimonious than invoking emotion states in animals. So we run into a logical inconsistency if we try to use Occam's razor to argue both *against* the existence of emotional feelings in animals *and* to explain the similar effects of anxiolytic drugs in both animals and humans. In the end, it seems simplest to argue that animals and humans do share common emotion states whose neural substrates are the targets of anxiolytic drugs, and that these states can also be accompanied by feelings in humans (figure 5.3A) (Fanselow & Pennington 2018).

But if this view is correct, then why do so many experimental psychiatric drugs that initially appeared effective in animal models (rats, mice) fail to show efficacy when tested in humans? There are several reasons. First of all, rodent and human physiology *are* different, not just in the brain, but in the rest of the body as well. Many drugs that have cured cancer in mice have failed when tested in humans; however, the new immuno-therapies for cancer that are proving so powerful in humans were developed based on basic studies of immune system function in mice. Second, the assays (tests) that are used for evaluating psychiatric drugs in rodents often use behaviors that humans do not display—for example, tail rattling, a defensive action displayed by rodents when they are threatened. So, it would not be surprising if a drug that blocked tail rattling in mice had no effect in humans—who lack tails. If we could obtain clearer evidence for a neural signature of emotion states in humans, then we could test whether similar signatures might be seen also in animals. If so, then

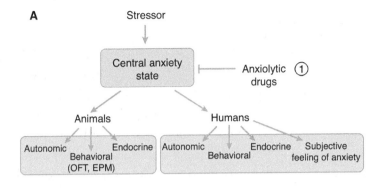

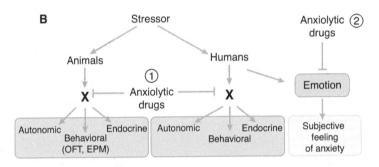

FIGURE 5.3. Two alternative explanations for the action of anxiolytic drugs. (A) Anxiolytic drugs quell a central emotion state of anxiety, which is common to humans and animals. In this view, anxiety is expressed in both species by common autonomic, behavioral, and endocrine responses; in humans, it is additionally accompanied by subjective feelings (which may be present in animals also but are unknown and therefore not indicated here for this reason). (B) Alternative model in which emotions are attributed to humans but not to animals. In order to explain the common effects of anxiolytic drugs in animals and humans, while adhering to this view, it is necessary to invoke at least *two* distinct mechanisms of anxiolytic drug action in humans: one (B ①) inhibits a process or state ("X") that is not an emotion, but that controls common physiologic responses to stressors in humans and animals; a second mechanism, unique to humans (B ②), inhibits the emotional component of anxiety. Model (A) is more parsimonious than model (B) in explaining anxiolytic drug action, but some consider model (B) more parsimonious than model (A) because it does not attribute subjective feelings to animals. OFT, EPM: Open field test, elevated plus maze (two behavioral tests used to measure anxiety in animals; see figure 6.6).

we could test whether anxiolytic drugs act on the same circuits in animals as they do in humans. In the long run, it may prove possible to evaluate new psychiatric drugs based on their effects on specific patterns of brain activity rather than behaviors, first in animals and then in humans.

BOX 5.4. **Optogenetics and genetics in monkeys.**

Applying optogenetics (box 5.1) in nonhuman primates (NHPs) faces several challenges not faced in its application to mice. First and foremost, because the NHP brain is so much larger than the mouse brain, there are major limitations on the relative proportion of brain volume that can be illuminated using conventional optic fibers, or virally infected using conventional adeno-associated viruses (AAVs), in comparison to the mouse. (Likely for these two reasons, early efforts at optogenetics in NHPs succeeded in evoking local spiking activity, but no behavioral responses.) Second, although germline transgenesis avoids the spatial limitations imposed by intracranial viral infusion, the generation of transgenic NHPs is both costly and lengthy (due to their long gestation time). Third, with a few exceptions (see below) we lack cell type-specific genetic regulatory elements (a.k.a. "promoters") that are small enough to fit in AAVs (the most common type of viral vector) for expressing opsins in specific classes of neurons. Fourth, validating the site and specificity of expression of opsins following optogenetic experiments is essential, but currently requires sacrificing the animal; NHPs are valuable, and investigators often use the same animal for many months or even years of experimentation, therefore they are loath to sacrifice such a precious animal simply for histological verification.

Fortunately, some of these obstacles are slowly but steadily being overcome. For example, new technologies have recently been developed for delivering illumination to the NHP brain over surface areas >100 x that of a conventional optic fiber, allowing activation of much larger volumes of brain tissue (~10 mm^3; Acker et al. 2017). And several recent papers have reported using cell type-specific promoters to express channelrhodopsin-2 in specific cell types in NHPs, such as dopaminergic neurons of the ventral tegmental area, or cerebellar Purkinje cells. (However, the number of different cell type-specific promoters available for use in NHPs, or in rodents for that matter, is currently very small.) The problem of delivering sufficient volumes of virus to infect a large enough volume of brain tissue may soon be solved using AAV "serotypes" (viruses of particular cell- or tissue-targeting specificities) that can cross the blood-brain barrier, so that they can potentially be delivered intravenously rather than via intra-cranial injection. Furthermore marmosets, which are more favorable for transgenesis because of their shorter generation time and tendency to produce twins, are being developed as an alternative to rhesus-macaque monkeys for genetic experiments. Such germline transgenic approaches in monkeys are being further enabled by the application of so-called "CRISPR/cas9" technology, which greatly increases the efficiency and specificity of stitching pieces of foreign DNA into monkey chromosomes. An important underdeveloped area, highly relevant to the topic of this book, is the establishment of sophisticated, quantitative assays of emotional behaviors in NHPs, which are currently extremely limited. Altogether, the prospects for applying optogenetics and chemogenetics for the study of emotion in NHPs seem considerably brighter than they did just a few years ago. However, the high cost of maintaining NHPs, the requirement for large numbers of animals in order to breed transgenic lines and the need to sacrifice animals for histological analysis, together with the stringent regulations imposed on NHP experimentation in the United States, will likely restrict the application of these approaches to a few well-funded centers, particularly in Asia, for the near future.

BOX 5.5. **Optogenetics in humans.**

Will optogenetics be used to treat psychiatric and emotional disorders in humans? Possibly, but it is doubtful that this will happen anytime soon, for several reasons. First (and foremost), it requires genetic modification of neurons in the brain, in order to express channelrhodopsin or halorhodopsin (or other opsins). Because germline transgenesis is currently banned in humans, viral vectors would have to be used to deliver the opsin genes. There is a high regulatory barrier for such a therapeutic modality set by the FDA, because the opsin genes are not human genes, but rather foreign genes from micro-organisms, which could prove toxic or could trigger an immune response. It may surprise you to know that it wasn't until October 2017 that the FDA finally approved the first viral vector-based gene therapy for a brain disorder (adrenoleukodystrophy, or ALD), using a modified HIV virus to deliver the gene. And in that case, the gene wasn't even being directly delivered into the brain; rather it was just being used to replace a defective gene in bone marrow cells, which then produce genetically "cured" immune cells that circulate through blood vessels in the brain.

Importantly, the regulatory justification for this type of gene therapy is very different from that for optogenetics: in the case of ALD, a defective gene is being replaced, without which the afflicted children will die. In the case of optogenetics, there is no "defective gene" being replaced; rather a foreign gene (the opsin) is being inserted into neurons to enhance or decrease their firing rate, in the hopes of ameliorating symptoms of a disorder like depression or anxiety. The regulatory bar is higher when the benefits are not clearly lifesaving, and the associated risks are higher. Second, the treatment (as currently practiced) would require viral injections into the brain and implantation of optic fibers (hardware), a highly invasive procedure. Third, we currently lack the ability to direct the expression of opsins to specific cell types in the human brain, not only because we do not yet have a census of cell types in the human brain, but also because we lack the necessary genetic "addresses" of those cell types ("promoters"). Fourth, even if we had non-invasive methods for viral delivery and cell type-specific promoters to target specific human cell types, optogenetics would still require the permanent implantation of optic fibers deep into the brain, and portable laser systems under closed-loop control to deliver light pulses for stimulation – in other words, a great deal of portable hardware including fragile components that could easily break. That said, clinical trials are planned to test the use of optogenetics in the eye to treat certain retinal degenerative disorders. It is more likely that chemogenetic approaches to manipulating neuronal activity (e.g., DREADDs; cf. box 5.1 and glossary) will be used to treat neurological and neuropsychiatric disorders, before optogenetic methods are used. Indeed, clinical trials are planned in the UK to test the used of inhibitory DREADDs to reduce seizures during epilepsy, by silencing cortical neurons at the seizure focus.

Summary

- In this chapter, we discussed some of the general considerations underlying the importance of studying the neuroscience of emotion in animals. We described the kinds of questions that can be asked about emotion systems in animals and some of the new tools that are being used to address them, such as optogenetics. Most of these tools cannot be used in humans.

- Optogenetics, despite frequent misunderstanding, is not exclusively a technique for artificially activating neurons, nor is it a technique for imaging neural activity. It can also be used to reversibly silence or inhibit specific populations of neurons. Together these tools, along with pharmacogenetic methods for activating or inhibiting neuronal cell types, allow one to test the necessity and sufficiency of specific cell populations for particular behaviors or internal states in animals.

- Optogenetic methods for manipulating neuronal activity are being complemented by new technologies for imaging activity in deep brain structures with cellular resolution, using genetically encoded calcium-sensitive fluorescent indicators. Conjoint application of this suite of technologies is making it possible to record and "play back" spatiotemporally precise patterns of neuronal activity associated with a particular behavior or internal state.

- We discussed the relationship between emotion states and other internal states such as motivation, arousal, and drive (MAD states). There is a general consensus that MAD states can be studied in animals, but some researchers define "emotions" as subjective states that can only be studied in humans.

- We discussed the strengths and limitations of testing drugs for human psychiatric conditions in animal models and argued from Occam's razor (principle of parsimony) that it is more likely than not that there is to be a common mechanism through which anxiolytic drugs quell symptoms of anxiety in humans and other mammals.

CHAPTER 6

The Neuroscience of Emotion in Rodents

If we want to study the neurobiology of emotion states in animals, we need to have a way to elicit those states. *Emotions*, as the word's etymology implies, are fundamentally *reactive* in nature; they involve movements (behaviors) triggered by stimuli in the external world as well as the internal representation of stimuli and behaviors. As we already noted, such internal representations can also be accessed purely from memory or during dreams, something we humans frequently experience, but about which much less is known in animals. Most neuroscientists studying emotions in animals are concerned with understanding how an external stimulus elicits an emotional response, because external stimuli offer better experimental control. By contrast, many studies in humans (see chapter 8) rely on internal representations of stimuli (for example, asking subjects to remember emotional events in their lives).

While studies of emotion in animals typically use experimentally controllable stimuli as the inducers of an emotion, a further choice then comes down to whether one wishes to elicit the emotion using a stimulus to which the animal shows an *innate* pattern of reactivity, based on its evolution and inheritance, or using one to which the animal has *learned* to respond, for instance, through prolonged training (box 6.1). Each approach has its advantages and disadvantages. In the following sections, we will describe what has been learned, and what is still unknown, about the neurobiology of learned and innate emotional responses using the examples of conditioned and unconditional fear, respectively, focusing here on studies in rodents and then (chapter 7) in the vinegar fly *Drosophila*. (Once again, we use the term *fear* to refer to an internal emotion state by way of example, without any necessary attribution of conscious awareness or subjective feeling to the animal.) We also discuss related emotional behaviors including aggression, which in humans would be accompanied by subjective feelings such as "anger" or "rage." These are valuable exemplars in animal studies of emotions, but they only begin to scratch the surface. They also raise important open questions: do the themes gleaned from these examples apply also to social emotions? Also to positively valenced emotions? We do not provide anything close to a survey of all the different emotion systems that have been investigated in animals, instead choosing to give the reader a glimpse of model systems in which the details have been best worked out to date with modern methods.

BOX 6.1. Innate versus learned freezing behavior.

A frequent source of confusion, which arises in the discussion of innate (or "instinctive") versus learned emotional responses, is whether the *response* is learned, or whether it is associations with the *stimuli* that are learned. Freezing and flight are commonly viewed as instinctive defensive behaviors; animals do not have to learn how to perform them. However, the *stimuli* that elicit these behaviors can either do so "innately," that is, without the need for training, or "conditionally," meaning only after training. For example, mice do not freeze if exposed to a 2 kHz tone, but will do so if that tone is repeatedly paired with a shock (figure 6.1). By contrast, freezing and flight can be evoked by certain specific visual stimuli (e.g., an overhead looming disk, figure 2.5) in naive mice, upon first exposure.

However, it has been argued that *all* freezing behavior observed in the laboratory is conditional. According to this view, the unconditioned response (UR) to a foot shock (the unconditioned stimulus [US]) is the activity burst; the freezing that typically follows this burst after the shock is terminated is a *conditioned* (or "conditional") response evoked by the context in which training occurred (Fanselow 1980). Evidence cited to support this view is that if the foot shock is delivered immediately after the animal is introduced into the training chamber—before it has had time to investigate and create an internal representation of the context—no postshock freezing is observed. Again, however, this does not mean that the animal has to learn *how* to freeze, rather it means that freezing under these conditions is evoked as a consequence of a learned association between the context (training chamber) and an aversive event (foot shock).

An extreme generalization of this view is that there is no such thing as "innate freezing"—that every instance of freezing behavior reflects a learned association with the context. According to this reasoning, if an animal is observed to freeze in a given context, it is by definition conditioned, and should freeze whenever it is exposed to this context. But this prediction is inconsistent with recent experiments in which freezing was evoked in the animal's home cage by optogenetic stimulation of Sf1$^+$ neurons in the dorsomedial subdivision of ventromedial hypothalamus (VMHdm; see main text). Once the optogenetic stimulation was terminated, no further freezing in the home cage was observed, implying that conditioning to the context did not occur (Kunwar et al. 2015). These data provide evidence that artificial activation of certain brain regions can produce unconditioned freezing. As these regions are normally activated when animals are exposed to a predator and freeze, it seems reasonable to argue that the brain contains circuits that indeed can mediate unconditioned freezing in response to a predator.

Emotion, Fear, and the Amygdala

The most intensively studied neural structure involved in generating fear-like states is the amygdala, a complex structure containing at least twelve different subdivisions (see figures 2.1, 6.2). Moreover, it functions through very close interaction with other brain structures. One of these, the bed nucleus of the stria terminalis, participates together with amygdala nuclei in processing anxiety (see below and figure 6.6B). Studies of the amygdala have received so much attention, in the popular as well as the scientific press, that some people have come to equate fear with the amygdala, or even to equate emotion per se with activity in the amygdala. As we noted in earlier chapters, however, this view is oversimplified and overgeneralized; while the amygdala plays an undeniably important role in *learned* fear, it is far from the "seat of emotion" in the brain. In fact, Joseph LeDoux, and who carried out seminal work on the role of the amygdala in conditioned fear (LeDoux 1996; 2000), has now reinterpreted the amygdala as a structure that controls defensive behaviors (for example, freezing) but not "emotions" (as he now has redefined them [LeDoux 2012]): "*feelings* of fear or anxiety are not *products* of circuits that control defensive behavior" (LeDoux and Pine 2016) (italics ours).

Does this statement mean that the amygdala is irrelevant to fear? More generally, does it mean that circuits that control defensive behavior are necessary but not *sufficient* to produce feelings of fear or anxiety, or that they are not even *necessary* to experience such feelings? We would agree with the former but disagree strongly with the latter; as we shall see in chapter 8, there are patients lacking an amygdala who do not experience fear. For this reason, we feel that work on the amygdala is highly relevant to the question of how emotions are generated and represented in the brain. In fact, we (like others including Jaak Panksepp) argue that subcortical structures that control defensive behavior (i.e., amygdala, hypothalamus, and related structures) are likely to be essential to the encoding of corresponding central emotion states—whether the animal has subjective "feelings" of those states or not.

What Is "Fear Conditioning"?

Because older work on this topic has been covered in comprehensive reviews (Davis 1992; LeDoux 2000) and books (LeDoux 1996) we will be brief here and refer the interested reader to those earlier sources for further information. Ironically, the conditioned fear paradigm was developed not so much to study emotion per se, as to study associative learning. Fear conditioning represents a form of Pavlovian conditioning that one might call "emotional learning": the ability of the brain to learn to respond to a previously neutral stimulus with an emotional response (Fanselow 1984). LeDoux no longer uses the term "fear conditioning" but rather calls it "defensive conditioning" (LeDoux 2012) to emphasize that (in his view) this conditioning does not produce an emotion. However, this is not a widely adopted term; indeed, at the time of writing this book, a PubMed search using "defensive conditioning" returned 69 hits, whereas "fear conditioning" returned 4,649 hits. Since we will argue, as have others (Fanselow and Pennington 2018), that such conditioning indeed produces an internal emotion state, and since fear is an emotion, we will continue to use the more common term "fear conditioning."

In this simple form of learning, an animal is exposed to a previously neutral "conditioned stimulus" (CS), such as a neutral tone, that co-terminates with an "unconditioned stimulus" (US) such as a foot shock (this protocol is called delay conditioning). After several such pairings, the CS evokes freezing, a conditioned defensive response (CR; Fanselow 1980) (figure 6.1; see also box 6.1). Remarkably, recent studies have shown that a rat given just a single day of training can exhibit a freezing response even when testing is delayed for a year, indicating the remarkable durability and adaptive value of this form of learning. In humans, this endurance of learned fear responses can under certain circumstances result in psychiatric disorders, such as post-traumatic stress disorder.

Importantly, the conditioned response does not consist simply of freezing behavior. Rather, it comprises a constellation of coordinated physiological (autonomic) and endocrine (hormonal) as well as behavioral responses (figure 2.1). These can be measured, for example, as increased heart rate (autonomic) or increased levels of stress hormones in the blood (endocrine). They also include other measurable changes such as decreased salivation, changes in respiration, increased urination and defecation, scanning and vigilance, increased startle in response to a loud noise, and grooming (Davis 1992). It is this coordinated, "global organismal response" (LeDoux 2012) that we argue reflects the triggering of an internal emotion state (remember, this was one of the properties of emotions we highlighted in figure 3.2). It is also important to note that the amygdala's role in fear conditioning can show quite different functional properties in very young animals (box 6.2). We focus only on emotions in adults in this book, but of course emotional development is a huge and important topic in its own right.

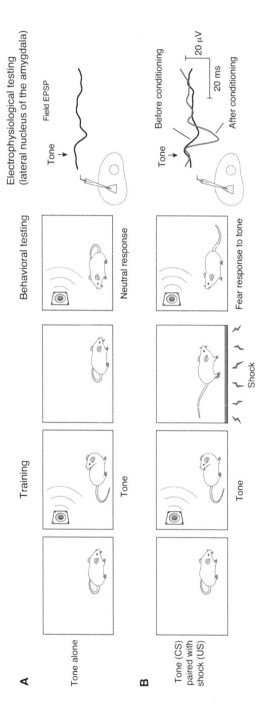

FIGURE 6.1. Pavlovian fear conditioning. (A) Unconditioned control animal exposed to a 2 kHz tone shows no defensive response to this stimulus. (B) If the tone is presented so that it coterminates with, or is immediately followed by, a foot shock, then the animal acquires a conditioned freezing response to the tone when tested the following day ("behavioral testing"). If one records from neurons in the lateral amygdala ("electrophysiological testing"), one finds that the neurons show a large change in response (compare black and red traces). Reproduced by permission from Kandel et al. 2013.

BOX 6.2. **Development of emotions.**

The development of emotions involves a complex interaction between genes and environment, between innately programmed mechanisms and learned associations. For example, there are innately expressed behaviors, such as smiling, that are seen even in human newborns. However, they are expressed in many circumstances, even during sleep. Such behaviors are, over time and with learning, incorporated into the expressions seen in full-fledged emotion states.

The emotional behaviors shown in young animals and in human babies support the functional approach we have stressed. Human infants strongly guide their emotional behaviors depending on the social context: whether somebody is watching, who is available to help them, and what feedback they get from people's behavior. This requires a lot of learning, and baby humans and animals are prodigious learners right from birth (indeed, even before birth).

Given that infants are not yet able to defend themselves and cope with many environmental challenges on their own, their emotional behaviors often serve a different functional role than in adults. Indeed, it makes sense that they would not show adult emotions, because these would mostly be useless to them. Trying to run away in fear won't work if you can't run; trying to bite a predator won't work if you don't have teeth yet. Consequently, the emotional behaviors of infants are instead geared toward getting adults to cope with these challenges on behalf of the infant. The emotional behaviors are still instrumental and goal directed, but the proximal goal is the parent, not the environmental challenge.

An interesting example of an emotion that predominates in infancy is attachment (Sullivan and Wilson 2017). Aspects of its mechanisms involving the amygdala have been worked out in considerable detail (Sapolsky 2009). Remember that the amygdala plays a key role in Pavlovian fear conditioning: if a stimulus is paired with an aversive outcome (e.g., pairing a particular odor with electric shock), animals will learn to avoid that stimulus. However, most studies on this topic were done in adult animals and humans, and it turns out that early in development things work quite differently.

Up until about ten days of age, rat pups are completely dependent on their mother; later, they begin to move around independently. When odor-shock conditioning is tested in older rat pups, we find the usual emotional learning: they learn to avoid the odor that was paired with the shock. However, young pups (< 10 days old) show the opposite effect, and instead show an attraction to the odor that was paired with the electric shock! One interpretation that has been given to these results is that they reveal a mechanism whereby very young animals will form attachments regardless of the circumstances. This makes some functional sense, since any attachment at all may be better than none, given that the pups are so completely dependent on an adult for their survival. This same mechanism may also explain sad cases of human infants or children that were abused for years but ended up forming strong attachments to the people who abused them.

There is a final important point made by developmental studies. Not only do the specific emotion states seen at particular developmental periods serve often unique functions (such as separation anxiety or attachment), but there is also a general differentiation with development. Human infants exhibit behaviors and facial expressions related to positive, affiliative situations, and related to distress. But these are quite general, undifferentiated emotion categories, from which specific emotions like fear, anger, or sadness emerge only later, and likely require a social context.

Historical Work on Fear Conditioning Circuitry

Classical studies of the neural circuitry of fear can be conceptually divided into those that have addressed the mechanisms of conditioning (acquisition), and those that have investigated the implementation (expression) of the conditioned response. It is the latter that is most relevant to the topic of this book; ironically, it has received far less attention than the former, reflecting the provenance of the field in experimental psychology and learning theory. Indeed, "fear conditioning" (and its insect counterpart; see chapter 7) is arguably the best experimental model available to study the mechanisms of associative learning per se, in any part of the brain.

Experimental lesions and electrical stimulation have indicated that the amygdala is involved in both the *acquisition* and the *expression* of conditioned fear. Only about three to five of the amygdala's twelve nuclei (depending on their definition) are involved in fear conditioning: the lateral nucleus (LA), baso-lateral nucleus (BLA), and central nucleus (CE, or CeA) (figure 6.2C and figure 6.6C). The rest are involved in processing olfactory information (for example, the cortical amygdala), and in mediating social and reproductive behaviors (medial amygdala; each of these regions have further subdivisions). Nevertheless, because of the disproportionate attention and popularization focused on the role of the amygdala in fear conditioning, many people have assumed that any neuroimaging studies reporting activity in the human "amygdala," unless otherwise specified, refer to activity in LA, BLA, or CeA—but this inference is incorrect. In fact, until relatively recently most fMRI scanners did not have the resolution to distinguish activity in any of these different nuclei, and even now this remains problematic (see chapter 8). This issue is even more important when one realizes that there are many different neuronal cell types in the amygdala that all subserve somewhat different functions, none of which can be resolved at all by even the most powerful fMRI scanners (figure 6.4).

An extensive body of work involving stimulation, lesions, pharmacological inactivation, and electrophysiological measurements has led to a "standard model" in which the acquisition of fear conditioning is mediated by the LA and/or BLA, while expression (of the CR) is mediated by the CeA (Davis 1992; LeDoux 1996, 2000). According to this view, the pathways that convey the CS (cue) and US (shock) converge in the LA, where they strengthen synaptic responses to the CS (figure 6.2A). This information is then relayed to the CeA, which is the final common pathway for output from the amygdala (cf. figure 2.1). However, more recent evidence indicates that this view is oversimplified. First of all, the CeA is not a unitary structure, but rather consists of at least three different subdivisions: the lateral (CeL), the medial (CeM), and the

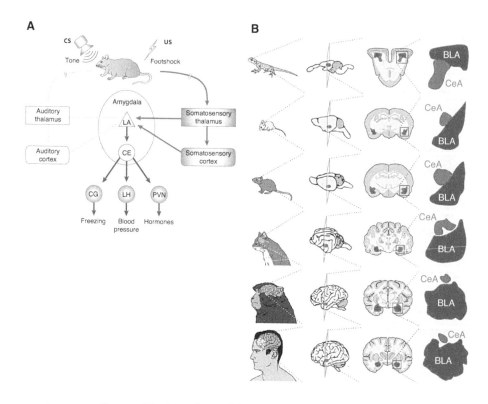

FIGURE 6.2. Fear conditioning and amygdala nuclei. *Left*: Simplified schematic illustrating basic components of fear conditioning circuitry. The BLA is not included in this diagram. CE, central amygdala (equivalent to CeA). *Right*: The main nuclear complexes, basolateral (BLA) versus central (CeA) amygdala, are found across all vertebrate animals. Other abbreviations: CS: conditioned stimulus; US: unconditioned stimulus; CG: central gray (same as periaqueductal gray); LH: lateral hypothalamus; PVN: paraventricular nucleus. Reproduced with permission of Macmillan Publishers Ltd.: *left* from Medina et al. 2002; *right* from Janak and Tye 2015.

capsular (CeC) (figure 6.3). Recent studies of the pathway conveying the US signal to the amygdala indicate that the US and CS pathways also converge on neurons in CeL. Thus the traditional view that LA is the principal site of learning, while CeA is simply an output structure that mediates the expression of conditioned fear, is being revised. The internal circuitry within the amygdala is more complicated than that.

If one considers (as we do) that the diverse organismal responses evoked by the CS following acquisition are indicative of an underlying central emotion state, then to the extent that CeM organizes these responses, it should play a role in the implementation of this state. Work in the late 1980s and 1990s established that the CeM coordinates behavioral, autonomic, and endocrine components of the CR via divergent projections to different downstream brain regions, each of which is thought to play a more specific role (Davis 1992; LeDoux 2000) (figure 2.1). Whether these divergent projections derive from common or distinct neuron types is not yet clear. Perhaps counterintuitively, these CeM outputs are made by inhibitory projection neurons, which disinhibit target cells in downstream structures such as the periaqueductal gray (PAG) (figures 6.2 and 6.3A). These cells in turn control premotor circuits that regulate freezing or flight behavior.

Thus, the data are consistent with a view that CeA plays a role in *coordinating* different aspects of a conditioned emotion state, in a *hierarchical* manner. So here is one example where we have found in the brain a neural implementation of one of the properties of an emotion state we discussed in chapter 3—the property of global coordination, achieved here, at least in part, via a hierarchical scheme. However, there have been few if any studies that have systematically investigated whether the CeM encodes any of the other properties of emotion states discussed in chapter 3 (for example, scalability, persistence, generalization), and if so, how. It is possible that different properties, or subsets of properties, of an emotion state (the different entries in figure 3.2) might be implemented by distinct neural components, almost certainly including structures outside the amygdala.

Recent Work on Fear Conditioning Circuits:
Cellular Heterogeneity in the Amygdala

The last decade has seen a revolution in the development of powerful new tools for investigating the organization and function of neural circuits, including optogenetics, pharmacogenetics, and calcium imaging, as we already noted (see boxes 5.1–5.3). The power of these tools derives from the fact that they can be genetically targeted to specific neuronal subpopulations in mice (box 5.3; unfortunately, these tools are still not widely applicable in rats, let alone in primates, see box 5.4). The application of these tools to the fear conditioning system has produced an avalanche of new information about these circuits, at an unprecedented level of cellular resolution and specificity (Janak and Tye 2015). Integrating these results from mice with historical studies of fear conditioning circuitry, which were almost exclusively performed in rats, requires an assumption of cross-species neuroanatomical and functional homology. This assumption is probably valid in most, but not necessarily all, cases.

Although the field is still relatively young, the most important general conclusion to emerge thus far is that amygdala cell populations are far more functionally heterogeneous than previously anticipated. In some cases, even subdivisions of amygdala subnuclei have been shown to contain a mixture of multiple cell types, which can exert opposite effects on fear conditioning. For example, in one of the first studies of its kind, the lateral subdivision of the central nucleus (CeL) was shown to contain at least two molecularly distinct subpopulations of GABAergic neurons (neurons that use the neurotransmitter gamma-aminobutyric acid [GABA]) distinguished by the expression of a specific molecule, protein kinase C-delta (PKCδ). Electrophysiological recordings in live mice, together with genetically targeted manipulations of neuronal activity (boxes 5.1–5.3) showed that neurons that express, versus those that do not express, this molecule (PKCδ⁺ and PKCδ⁻ neurons, respectively) exert antagonistic influences on fear conditioning, with the former type of cells inhibiting, and the latter disinhibiting, CeM output (figure 6.3A). Moreover, these CeL subpopulations reciprocally inhibit one another, creating the potential for winner-take-all circuit dynamics within this structure.

Optogenetic manipulations have confirmed that CeL circuits play a role in the acquisition of conditioned freezing, indicating that this plasticity is distributed across different amygdala subnuclei and not restricted to the LA/BLA as originally thought (Janak and Tye 2015). Consistent with this view, a US pathway for fear conditioning that conveys information to the amygdala via CeL has recently been identified, involving yet another specific subpopulation of neurons in another brain nucleus (the external lateral parabrachial nucleus, which in turn receives ascending input from spinal neurons activated by foot shock (Han et al. 2015)). Optogenetic activation of these CeL neurons

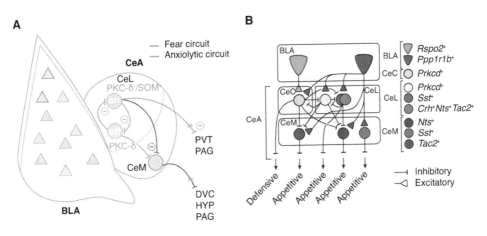

FIGURE 6.3. Neuronal subpopulations in the central amygdala. (A) Different CeA neuronal subpopulations can be identified according to whether they differentially express certain genes (e.g., protein kinase C-delta; PKCδ, or somatostatin; SOM). These different populations can exert opposite direction effects on behavioral responses. "Off" cells inhibit fear responses; "on" cells promote fear responses. Purple cells are unidentified CeM output cells. PVT, PAG, DVC, HYP are regions to which the CeA projects. Colored triangles in BLA denote different cell types that might project to cells of the corresponding color in CeA. Reproduced with permission from Janak and Tye 2015. (B) More recent studies identify nine different neuronal subtypes in different subdivisions of CeA, using additional molecular markers (legend on the far right). Note that several CeM cell types promote appetitive behaviors. Reproduced with permission from Kim et al. 2017. These examples illustrate the complexity of central amygdala circuits, and the importance of being able to identify, and manipulate, brain function at the level of specific cell types so as not to mix circuit functions.

produces unconditioned freezing, and is sufficient to serve as a US for conditioning to a tone.

There are likely to be several further neuronal subpopulations involved in fear conditioning. Indeed, more recent studies by Susumu Tonegawa and colleagues have revealed evidence of as many as seven distinct cell types within the different subdivisions of the CeA, some of which are involved in appetitive behavior and some of which are involved in fear conditioning (Kim et al. 2017). These studies have shown that even CeM contains neurons that control appetitive behaviors (figure 6.3B). In general, the field is in a state of rapid growth, and there is likely to be further revision of these models over the next few years as more data and new methods become available. A critical step will be to investigate the patterns of activity of different CeA subpopulations during behavior, using tools such as calcium imaging (box 5.2).

Cellular Heterogeneity in the Basolateral Amygdala:
Implication for Theories of Emotion

One of the most striking recent examples of cellular heterogeneity in the amygdala is the discovery that the BLA contains distinct, inter-mingled populations of neurons that respond differentially to intrin-sically rewarding versus aversive stimuli. BLA neurons activated by rewarding stimuli, such as nicotine, project to the nucleus accumbens (NA), a region known to be involved in reward processing (figure 6.4B, green); optogenetic activation of these neurons supports positive re-inforcement. BLA neurons activated by aversive stimuli, such as foot shock, project to the CeA, and are necessary and sufficient for aversive conditioning (figure 6.4B, red) (Namburi et al. 2015). These import-ant data identify distinct subsets of neurons that represent positively and negatively valenced emotional stimuli, and which are interspersed throughout BLA without any obvious anatomical demarcation between them. It is interesting to note that similarly interspersed populations of neurons responding to aversive or appetitive stimuli have been found in monkeys—in this case, on the basis of electrophysiological response properties rather than on optogenetic activation (Paton et al. 2006).

These data have important implications for recent debates about the role of the amygdala in emotion. For example, it has been argued that emotions such as fear are not represented or encoded by specific circuits or brain regions, because activation of the amygdala can occur during both positive and negative emotions. This conclusion might be valid if the *same* neurons in the amygdala were activated during both positive and negative emotions. However, this conclusion is based on fMRI imaging studies in humans. The signal detected in fMRI experi-ments (figure 6.4A and chapters 8, 9) has neither the spatial resolution nor the cellular specificity to detect different subsets of BLA neurons as revealed by genetic marking and single-cell activity measurements in the mouse (figure 6.4B). The mouse studies reveal that it is not the case that the same neurons are activated by both positively and nega-tively valenced emotional stimuli: rather, the amygdala contains distinct neuron subsets that are specifically activated either during exposure to threatening stimuli or during exposure to rewarding stimuli. These

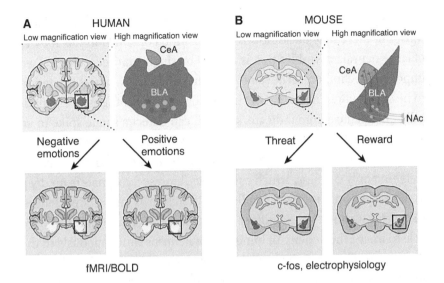

FIGURE 6.4. Resolution and cellular specificity of methods to measure amygdala activity in humans versus mice. (A) fMRI (*lower* panels) detects activation of the amygdala (BLA) equally during positive and negative emotions. This is because standard fMRI methods cannot distinguish activation of the same neurons versus activation of different populations of neurons. From this observation, it has been erroneously inferred that the amygdala has no specific role in fear. (B) Studies in the mouse reveal that the BLA contains different classes of neurons that promote either fear (red circles, aversive neurons), or reward (green circles, appetitive neurons). Studies measuring amygdala neuronal activity with single-cell spatial resolution and genetic specificity in the mouse show that these different neurons project to different targets (CeA or NAc, respectively) and are differentially activated, depending on whether the animal is exposed to a threat or a reward. These different classes of neurons are likely to exist in the human BLA (A, *upper*, high magnification view), although this has not yet been demonstrated. BLA: basolateral amygdala. CeA: central amygdala. NAc: nucleus accumbens, a region involved in reward processing.

different BLA cell types are likely to be present in humans as well (figure 6.4A, *upper right*), although this has yet to be proven. But if so, then the fact that the human amygdala can be activated during both positive and negative emotions (figure 6.4A, *lower*) does not mean that it plays no specific role in the implementation of emotion states; to the contrary, it plays specific roles in both. This example shows how important it is to be able to compare results from different methods, such as fMRI and optogenetics—comparisons that require a cross-species approach.

Another area in which there have been recent advances concerns the imaging of population activity—the firing of many hundreds of neurons—during and after fear conditioning. Electrophysiological recordings in the amygdala can detect and decode the activity of a few isolated neurons, but there may be additional information contained in the activity of large neural ensembles that is missed by single unit recordings. Two important technological developments have now allowed this limitation to be surmounted in mice. One is the development of sensitive genetically encoded calcium indicators, which can detect a single spike in an active neuron (box 5.2). The other is the development, by Mark Schnitzer and his colleagues at Stanford, of miniature head-mounted microscopes weighing only 2 grams, which can detect fluorescent signals from deep in the brain using tiny glass needles that function as microscopic lenses. Using this technology, Schnitzer and collaborator Andreas Lüthi were able to image the activity of hundreds of neurons simultaneously in the BLA during fear conditioning (Grewe et al. 2017). Remarkably, their observations showed that the ensemble representations of the CS and the US in naive (untrained mice) were initially quite different, but became more similar as the animals were trained to associate the CS with the US. Importantly, *decreases* in the activity of certain neurons were as important as increases in the activity of other neurons, to this reconfiguration of neuronal ensemble activity. These observations force a revision of the early view that fear conditioning only involves an increase in the activity of certain CS-responsive amygdala neurons (figure 6.1B) and illustrates the power of this new approach to reveal information contained in the activity of large, distributed ensembles of neurons. Many more applications of this technology to the study of emotion circuits in mice can be anticipated in the future.

What We Still Don't Know about the Amygdala and Fear Conditioning

Despite the remarkable and rapid (indeed, somewhat overwhelming) progress in the functional dissection of amygdala circuits involved in fear conditioning, there are still major gaps in our understanding of this important model system. Most of the progress that has been made revolves around the functional and anatomical analysis of genetically marked cell populations that mediate different aspects of the fear conditioning process. These studies have revealed a level of cellular heterogeneity and circuit complexity far greater than previously suspected, and have yielded important insights into the circuit-level coding of emotional valence. By contrast, much less is known about where and how other properties of the central emotion state evoked by a conditioned stimulus are subserved in the amygdala. Where and how are persistence, scalability, and generalizability (figure 3.2) encoded in the amygdala? Or indeed, are they encoded in the amygdala at all, or do they depend on other brain regions? Surprisingly little work has been done to date investigating whether CeM, the major output of the amygdala, instantiates any of the characteristic properties of emotion states that we listed in figure 3.2.

We also know relatively little about whether the systems driven by CeA output, in addition to sending signals to the brainstem and effector systems, also signal their activation to the neocortex, and if so through what circuits. Such ascending cortical projections could provide important substrates for those emotion properties that depend on further interactions with cognitive processes. It is known that the basolateral nucleus of the amygdala (BLA) projects to many regions of the cortex, and it may well be that the CeA also does so. Thus, we have achieved an increasing level of detail in our understanding of the amygdala and its circuitry as a learning machine, in large part due to important advances in methods like optogenetics. However, whether the amygdala is the exclusive or even the principal locus for the implementation of fear states in the brain remains unclear and controversial, and will require a more comprehensive investigation that looks more broadly across the whole brain.

Another area in which we are still relatively ignorant is the role in fear conditioning of neuropeptides, short proteins that bind to specific G-protein coupled receptors (GPCRs). While many of the studies already described have used neuropeptides as markers for specific classes of amygdala neurons, whether the functional influences of these neurons are mediated by the neuropeptides themselves, by co-released classical transmitters (typically GABA in the CeA), or by both, remains largely unexplored. For example, the peptide Tachykinin-2 (Tac2), which also marks a subset of CeA neurons, has recently been shown to promote the consolidation of learned fear. The roles of other neuropeptides expressed in the amygdala, which include enkephalins and dynorphins and many others, are still poorly understood. New advances in "genome editing" technology should make it possible to create null (complete loss of function) mutations in specific neuropeptide genes in specific classes of amygdala neurons, and advances in this field are likely in coming years. This work is also likely to lead to opportunities to develop new drugs for the treatment of fear- or anxiety-related disorders, which act on these neurotransmitter systems with greater specificity than existing drugs.

Regulation of the Amygdala by the Prefrontal Cortex

An important aspect of fear conditioning concerns interactions between the amygdala and the prefrontal cortex. It is widely assumed, based on lesions in humans (chapters 8 and 9), that the prefrontal cortex exerts a "top-down" inhibitory influence on subcortical emotion systems (box 3.4), and that dysfunction in such emotion regulation might contribute to disorders such as PTSD or depression. A well-studied manifestation of such top-down regulation is the phenomenon of "fear extinction," one of the few cases in which this regulation has been studied at the circuit level in rodents. Briefly, "extinction" refers to the phenomenon in which an animal that has been previously conditioned to respond to a CS gradually loses that response (CR) following repeated presentations of the CS. This has been shown to involve not simply "forgetting," but rather to reflect new learning: the animal learns that the CS no longer predicts the arrival of the US, but that it is now "safe" (analogous to learning to ignore "the boy who cried wolf!"). Extinction learning has been shown to involve descending excitatory projections from the medial prefrontal cortex (mPFC; in rodents the infralimbic cortex, IL), which suppress the expression of the CR in the amygdala. The mPFC is likely to have at least as many cell types as the amygdala, and current research has only begun to scratch the surface of this diversity, revealing distinct neuronal subclasses that project to different subcortical targets, including not only the amygdala but other regions as well.

How extinction works locally in the amygdala is not yet completely clear, but it is thought to involve recruitment of inhibitory neurons in a region called the intercalated cell mass (ITC), which in turn suppress output from the CeA (Quirk and Mueller 2008). Importantly, communication between the mPFC and the amygdala is not unidirectional; there are also outputs from the BLA that target specific mPFC neurons (as well as indirect connections via the thalamus). These studies begin to provide some insights into the complex, bidirectional communication between the prefrontal cortex and amygdala, another example of the distributed neural implementation of emotion states (Salzman and Fusi 2010).

It should be noted that interactions between the mPFC and the amygdala have received a disproportionate amount of attention in animal studies, because of evidence from human lesion patients (chapters 8 and 9), to the point that many people may think that these are the only (or primary) brain regions relevant to emotion. However, atlases of the mouse brain list over eight hundred anatomically distinct regions, and there are probably even more in the human brain. It is very unlikely that the mPFC and amygdala are the only ones involved in emotion; indeed, recent work has implicated other structures, such as the medial and lateral habenula, in aversive emotion states. There is plenty of work to do ahead!

Innate Defensive Behaviors and Emotions

In addition to responding to *conditioned* artificial stimuli (for example, the 2 kHz tone used in many fear conditioning experiments; see figure 6.1), animals will also exhibit *innate* defensive responses to certain specific, ecologically relevant stimuli, whose ability to evoke those responses has been selected over millions of years of evolution. These innate responses can be observed in laboratory animals the very first time that they experience the stimulus, without any prior training (Tinbergen 1951) (see box 6.1).

For example, a laboratory mouse exposed to an expanding overhead disk will exhibit a freezing or flight reaction the very first time it sees this stimulus (Yilmaz and Meister 2013) (remember that experiment we showed in figure 2.5). However, the response to such an innately aversive stimulus still exhibits considerable flexibility and is influenced by the context in which the stimulus occurs as well as by the species and genetic background of the test animal. For example, while outbred (more genetically variable) laboratory rats have been reported to freeze in response to trimethylthiazoline (TMT), a component of fox fecal odors, inbred C57Bl6 mice (a standard strain used for many behavioral and circuit analysis experiments that provides a uniform genetic background) do not exhibit true freezing in response to this odor, but rather display avoidance, risk assessment behavior, and periods of immobility (which can be distinguished from true freezing by a trained observer).

Responses to innately threatening stimuli also tend to be much more variable and heterogeneous than responses to conditioned stimuli, because the latter can be experimentally manipulated to evoke highly robust and repeatable responses by manipulating training parameters (for example, shock intensity, number of training trials, and such). All of these factors have made the study of the neurobiology of innate fear more challenging than that of the neurobiology of conditioned fear. This is somewhat ironic, since of course the innate responses are a prerequisite for the conditioned ones.

Historical Work on Innate Fear

A central question addressed in studies of unconditioned (innate) fear has been whether it uses the same circuits and pathways as those engaged by conditioned fear. This simple question has been more challenging to address than you might think. One reason is that the stimuli used to elicit innate responses have often been complex and multimodal (for example, intact predators), or of a different sensory modality (for example, predator odors) than those used to evoke conditioned defensive responses (typically auditory tones). Another reason is that the behavioral outcomes are often different (for example, risk assessment and avoidance of a predator versus freezing to a conditioned auditory stimulus). If the sensory inputs and behavioral outputs manipulated and measured in learned versus innate fear paradigms are different, there of course will be many differences in the respective neural circuits engaged—but those differences may not necessarily reflect differences in the encoding of innate versus learned fear. Ideally, to determine whether conditioned and innate stimuli share a common circuit representation of the internal state of fear, sensory stimuli of the same modality should be used to evoke innate and learned behavioral responses, and these responses should be the same. So far, this has been challenging to achieve.

Not surprisingly, therefore, historical studies of innate (unconditioned) versus learned (conditioned) fear have identified different brain regions involved in these two paradigms. For example, lesions of the LA and CeA indicated that these amygdala nuclei are required for conditioned but not unconditioned responses, whereas parts of the medial amygdala (MeA) and bed nucleus of the stria terminalis (BNST) have been shown, conversely, to be required for unconditioned but not conditioned responses (Rosen 2004) (figure 6.5A–D). However, this difference may simply reflect the fact that predator odors were often used to elicit unconditioned responses, and the MeA is well known to play a role in the detection of such olfactory stimuli; conversely, the LA processes input from the auditory thalamus, so its requirement for a response to an auditory CS reflects, at least in part, that sensory bias.

These studies have thus led to a prevailing but possibly erroneous view in which unconditioned and conditioned fear are processed by parallel and largely nonoverlapping pathways (Gross and Canteras 2012). More recently, a comparison of innate versus conditioned freezing evoked by different odors identified a cell population in the lateral subdivision of the amygdala central nucleus (CeL) that controls the balance between these behaviors, suggesting a common role for CeL in innate versus learned responses. While these interesting studies need to be replicated and extended, they give a flavor for how challenging it is to investigate this question, and the difficulty of designing experiments rigorously so as to allow "apples-to-apples" comparisons. Needless to say, this point applies in spades once we try to make comparisons between rodents and humans, from studies using different methodologies.

Do Circuits Mediating Innate Defensive Behaviors Contribute to Emotion States?

As mentioned earlier, some researchers argue that the amygdala (LeDoux and Pine 2016), and subcortical circuits in general (LeDoux 2015), control "defensive behaviors" but not emotions, in the sense of subjective feelings of emotions. According to that view, conscious experiences of emotions are instead instantiated in the cortex, not in subcortical structures like the amygdala (LeDoux and Brown 2017). An alternative viewpoint is that subcortical structures are directly involved in affective experience, in both animals and humans (Panksepp 2011b), in the sense that activity in these structures gives rise, indirectly, to activity in cortical structures that may underlie emotional feelings. While there are clear examples where subjective emotional feelings have been evoked in human patients by electrical stimulation of subcortical structures, those findings suffer from the limitation that the sites of electrical stimulation were poorly defined and do not exclude activation of the cortex through accidental stimulation of white matter (axons) close to the site of stimulation, and that the responses measured might be too nonspecific (for example, that they could just reflect generalized arousal rather than a specific emotional feeling) (LeDoux 2015) (chapter 8). However, with the advent of optogenetic tools and the ability to precisely perturb genetically defined neuronal subpopulations, it has become possible to revisit this question in animals—not by using the criteria of subjective feelings as a readout for the effects of stimulation on emotion, but by searching for specific response properties characteristic of central emotion states (chapter 3).

This approach has recently been applied to studying neurons in the ventromedial hypothalamus (VMH), a structure long known to be involved in mediating innate defensive behaviors (figure 6.5D, E). This nucleus in the hypothalamus is a key output in the so-called medial hypothalamic defensive system (Canteras 2002), which is thought to transform sensory inputs, detected by structures such as the MeA and posterior BNST, into behavioral outputs that are relayed through premotor nodes such as the dorsal periaqueductal gray (PAG) (Swanson 2005) (figure 6.5D). VMH is known to play a role in both predator defense behaviors and in social behaviors such as mating and aggression. In male mice, within VMH, these functions are segregated between genetically distinct neuronal subpopulations located in anatomically distinct subdivisions: predator defense is controlled by neurons located in the dorso-medial and central portion of VMH (VMHdm/c), which express the transcription factor Sf1/Nr5a1, while social behaviors are controlled by neurons located in the ventro-lateral portion of VMH (VMHvl), which express the type 1 estrogen receptor (Esr1) and the progesterone receptor (PR) (figure 6.5E). The expression of these specific genes by these two VMH subpopulations makes it possible to selectively manipulate their activity despite their anatomic juxtaposition, just like we saw earlier for specific subpopulations in the amygdala.

Pharmacogenetic silencing (box 5.1) or genetic ablation of VMHdm/c Sf1/Nr5a1⁺ neurons strongly reduced defensive responses (principally avoidance) to a predator. Conversely, optogenetic activation (box 5.1) of these neurons evoked defensive behaviors, including avoidance, freezing, and escape. It also caused increases in autonomic and endocrine function, as measured by pupillary dilation and serum corticosterone levels, respectively. Sf1⁺ neurons are also activated in the presence of a predator, as assessed by c-fos labeling (box 6.3). These data indicate that VMHdm/c Sf1⁺ neurons are necessary and sufficient for the expression of innate defensive behaviors, and that they are active during these behaviors under natural conditions.

BOX 6.3. c-fos and immediate early genes.

In rodents, patterns of neuronal activity associated with a particular behavior or state are often mapped using "immediate early genes" (IEGs) such as *c-fos* or *Arc*. IEG expression, which can be detected by antibody staining or in situ hybridization, is rapidly induced by elevated intracellular free calcium. Since neuronal activity increases intracellular free calcium (box 5.2), the induction of IEGs is often used as a surrogate marker of neuronal activation. Advantages of IEG mapping are that it has single-cell resolution (see figure 4.2), and can be used to systematically map patterns of activity across the entire brain. However, it also has certain limitations, like any technique. First, as commonly practiced it must be performed on fixed tissue; therefore, most contrasts between control and experimental conditions must be performed as between-subject rather than within-subject comparisons, unlike fMRI. (An exception is a modification of the technique called "catFISH," or

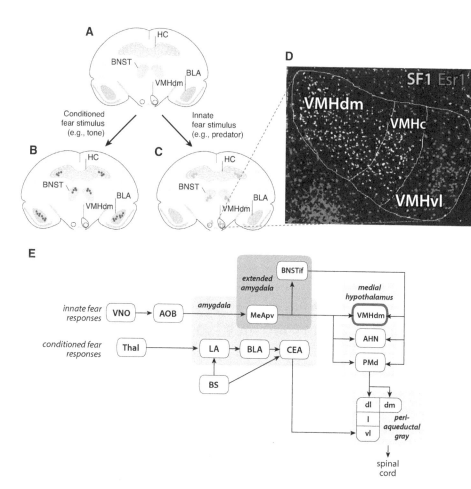

FIGURE 6.5. Distinct brain regions involved in different types of fear. (A) Schematic illustrating coronal section of a mouse brain indicating hippocampus (HC), bed nucleus of the stria terminalis (BNST), basolateral amygdala (BLA), and ventromedial hypothalamus (VMH). Note that these regions are compressed into a single plane for illustrative purposes. (B) Areas containing neurons (red dots) activated by a conditioned fear stimulus. (C) Areas containing neurons (green dots) activated by an innate fear stimulus. Different types of neurons in the BNST and HC are activated by the two types of fear stimuli. Cells and regions are not drawn to scale. (D) Section through mouse VMH stained with antibodies to SF1 (green) and Esr1 (red). Note the clear separation of SF1+ vs. Esr1+ neurons in VMHdm/c vs. VMHvl, respectively. (E) Schematic circuit diagram comparing pathways thought to mediate innate defensive response to predator odors vs. conditioned defensive responses. VMHdm (the dorso-medial subdivision of VMH) is outlined in red (see also D). VNO, vomeronasal organ; AOB, accessory olfactory bulb; MeApv, Medial Amygdala posterior ventral; BNSTif, Bed Nucleus of the Stria Terminalis interfasicular division; AHN, anterior hypothalamic nucleus; PMd, dorsal Pre-Mammillary nucleus; BS, Brain Stem; CEA, amygdala central nucleus; dl, dorso-lateral; dm, dorso-medial; l, lateral; vl, ventro-lateral.

cellular compartment analysis of temporal activity by Fluorescent In Situ Hybridization, which allows within-subject comparison of IEG expression under two different conditions spaced ~30 minutes apart.) Second, it has very low temporal resolution. Neuronal activity occurs on a time scale of milliseconds, yet IEGs typically take 60 to 90 minutes to reach levels of expression detectable by antibody staining (although expression can be detected within ~5 minutes by more sensitive methods). Therefore, the pattern of c-fos expression detected in a brain integrates all of the activity that occurred over a behavioral experiment, making it difficult to assign activity to specific actions. For example, c-fos expression detected following a 30-minute aggressive interaction between male mice or rats cannot distinguish neurons that were activated exclusively during the initial phase of social interaction (sniffing, investigation), or during fighting per se. Third, the expression of c-fos in a given brain region does not mean that output from that region is increased, although that is often mistakenly assumed to be the case. If c-fos activation occurred in inhibitory neurons, for example, output from that region might be decreased, whereas if it occurred in excitatory neurons output would be increased. Since c-fos expression alone cannot distinguish excitatory from inhibitory neurons, no conclusions can be drawn about the relationship between fos-activity in a given region and output from that region (a similar caveat applies to fMRI, as discussed in chapter 8). However, if c-fos staining is performed together with labeling for markers of excitatory or inhibitory neurons, it may be possible to resolve this issue (although most c-fos studies do not include this step). Fourth, IEG labeling is notoriously insensitive: only the most active (rapidly spiking) neurons are detected. Therefore, the absence of IEG labeling does not imply an absence of any activity at all. Finally, IEGs are not only induced by elevated intracellular free calcium, but also by other "second messengers" such as cyclic AMP (cAMP) which are commonly induced by neuropeptides and neuromodulators. Thus, detection of an IEG in a cell does not necessarily mean that the cell has been electrically stimulated; it may have been activated by a hormone or neuromodulator, instead. In the end, IEG labeling (like fMRI) is a good starting point; it doesn't tell you what the answer is, but it tells you where to look.

Importantly, these neurons appear to play a role in the encoding of some of the specific emotion properties that we listed in figure 3.2 (Kunwar et al. 2015; Wang, Chen, and Lin 2015). For example, optogenetic activation of these neurons induces an intrinsically negative *valence*, since it causes mice to avoid a chamber in which this activation is performed, in a so-called real-time place avoidance assay. It also causes a *persistent internal state*, as measured by persistent poststimulation avoidance responses. The behavioral effects of Sf1$^+$ neuron activation in VMHdm/c also exhibit *scalability*: lower intensities or frequencies of stimulation trigger avoidance, intermediate intensities trigger freezing, and high levels trigger escape behavior/activity bursts.

With respect to the property of scalability, the "rank order" of behaviors evoked by increasing activation of VMHdm/c Sf1$^+$ neurons resembles the order in which one would predict to see these behaviors emerge with increasing proximity of a threat, according to predator imminence theory (see figure 2.4). Conceivably, as a predator approaches (and attack grows more imminent), increasing activity of Sf1$^+$ neurons could cause the animal to switch from freezing (to evade detection) to flight (to evade capture). These observations suggest that the brain may *integrate* the cumulative level of Sf1$^+$ neuron activity over time to control a behavioral decision, similar to the drift-diffusion models we introduced in chapter 3 (figure 3.4). In this way, a categorical distinction in behavior may emerge from a continuous increase in the level of fear, which is translated into behavioral switches through non-linearities at downstream sites. Alternatively, as others have argued based on fMRI studies in humans, different systems may be activated when a threat is distal versus proximal, such as the BNST and the amygdala, respectively.

There are several other features of Sf1$^+$ neurons that provide a nice fit with many of the emotion properties we listed in figure 3.2. For instance, the *multicomponent* effect of activating Sf1$^+$ neurons (parallel changes in behavioral, autonomic and endocrine measures) is evidence that they play a *coordinating* role, similar to CeA (see figure 2.1). Sfl[+] neuron activation also produces *priority over behavioral control* in that it interrupts ongoing consummatory behaviors in males, such as mating, fighting, or feeding. Finally, activation of these neurons can serve as an unconditioned stimulus for conditioned place avoidance (Kunwar et al. 2015), indicating that they can play a role in *learning*. This last observation provides evidence to refute the idea, dating back to the 1940s, that the hypothalamus does not play a role in emotion because it cannot serve as an unconditioned stimulus for emotional learning—a weak conclusion that was based on negative experimental results.

Together, these data suggest that activation of VMHdm/c Sfl[+] neurons is sufficient to produce an internal defensive emotion state, whose behavioral expression exhibits many of the features of emotion states that we listed in figure 3.2. Whether the activation of these neurons causes the mice to feel "fear" in the way that we subjectively experience it as humans is not something we can answer yet—and, you will have noticed, is not a property of emotion states that we listed in chapter 3. That does not mean that the animals do not have these feelings; rather, it means that scientists do not have to attribute such feelings in order to study the neural encoding of a central emotion state. Moreover, the genetically targeted methods used in mice overcome the aforementioned criticisms leveled at studies reporting emotion states evoked by electrical stimulation of subcortical structures in humans (LeDoux 2015) because of their greater cellular specificity. To the extent that we view "feelings" in humans as the conscious experience of an internal emotion state, subcortical circuits that generate such states could well be an essential component or substrate of those subjective feelings (see also Panksepp 2011b). This is not to claim that activity in subcortical structures is *sufficient* to encode a subjective feeling in a human being; however, there are data in humans suggesting that it may well be *necessary* for the subjective experience of emotion states (see chapter 10).

The Hypothalamus Is Not Just for Eating and Drinking

As already reviewed, there is strong evidence that Sfl[+] neurons in VMHdm/c, a region of the hypothalamus, can generate a defensive internal emotion state. If one considers the hypothesis that such states are in turn a necessary substrate of conscious emotional experiences in humans (a hypothesis we find plausible), then this poses a challenge to prevailing views of the hypothalamus. The hypothalamus is most intensively studied as a brain region that controls homeostatic functions, such as feeding, drinking, and thermo-regulation (LeDoux and Damasio 2013; Sternson and Eiselt 2017). However, the results from animal studies reviewed above suggest that some regions of the hypothalamus may play a role in coordinating central emotion states as well. Moreover, the defensive emotion state evoked by VMH stimulation bypasses the amygdala (which is connectionally upstream, that is, providing inputs to the VMH). Consistent with this, evidence from humans indicates that certain fear states can be subjectively experienced in the absence of an amygdala (see box 8.2). Together, these observations underscore the point that the state of fear is not instantiated in a single locus in the brain, but rather is instantiated in a series of distributed and interconnected nodes, or organizing centers, that are engaged to different extents in different situations, depending on the animal's context, experience, and needs (figure 6.5). Only by dissecting experimentally the individual components of the system can we fully understand the mechanism by which the system ultimately links stimuli to internal states and behavior.

Distributed versus Localized Emotions in the Brain

It has been argued, for instance by Lisa Feldman Barrett (Feldman Barrett 2017), that there are no specific brain regions involved in the implementation of an emotion such as fear, because many different areas are activated during the experience of fear in humans, as assessed by fMRI imaging (and most of these regions are also activated by other emotions—as well as by nonemotional processes). But this is a bit like arguing that because a person tracked by a GPS transmitter from a satellite can be seen in several different places, according to when the data are collected, that the person has no permanent residence. Similarly, animal studies teach us that neurons in multiple brain regions can be activated during a particular type of fear (for example, the hippocampus and the amygdala during conditioned fear), and that different areas can be activated during different types of fear (for example, BLA and VMH in conditioned versus innate fear) (figure 6.5). Most importantly, optogenetic stimulation experiments show that activation of specific neurons within these specific regions is able to produce behavioral, autonomic, and endocrine features of a defensive emotion state, and that silencing or genetic ablation of these neurons abrogates emotional responses. Together these observations argue that emotion states are instantiated in distributed networks, made up of multiple localized nodes containing specific cell types involved in generating these states (figure 6.6 B,C). In short, the claim that there is no neuroscientific evidence for specific brain regions that control emotion states is both factually and logically flawed.

Note, however, that the causal evidence for specific emotion systems becomes weaker if one insists that emotions can only be studied in humans, because neuroscience studies in humans are largely correlative, and correlation does not imply causation (chapter 4). The fact that the argument against functional localization of emotion systems requires the wholesale dismissal of animal studies indicates, to the contrary, the importance of animal experiments to this debate.

Anxiety

"Anxiety" is generally not included among a list of the "primary" emotions (chapter 1), perhaps because it is an anticipatory, rather than a reactive, state that does not need to be triggered by any specific external stimulus. However, anxiety is clearly an internal affective state that shares many of the properties of emotion states we have discussed earlier, and one that has been studied intensively. One way of thinking about it is as a particular type of fear that is elicited when a threat is still far away, and the probability of needing to escape or defend oneself is relatively low (see box 2.3). It is expressed not only by measureable behavioral changes but also by physiological and endocrine responses similar to those measured in assays of fear. In humans, anxiety can have a major impact on cognitive processes such as decision-making.

Anxiety disorders are among the most prevalent psychiatric conditions: generalized anxiety disorder (GAD) afflicts close to seven million adults in the United States, over 3 percent of the population. Yet current treatments for anxiety are unsatisfactory, largely because of the side effects of the drugs. For example, many people cannot tolerate benzo-diazepines (drugs like Valium) during the day because of their strong sedative effect. SSRIs ("selective serotonin reuptake inhibitors," drugs like Prozac) are also used to treat anxiety, but these have serious side effects as well, and many individuals do not respond to or tolerate them. An understanding of the neuroscience of anxiety will be critical to the development of better treatments.

Scientists working on anxiety in rodents (rats or mice) typically use different behavioral assays than those used in fear conditioning. These assays include the open field test, which measures the animal's avoidance of open spaces; the elevated plus maze, which measures the animal's decision to enter raised platforms with or without walls; the light-dark box, which measures the animal's avoidance of brightly lit places; and other tests such as marble-burying, hole-board, and novelty-suppressed feeding, which measure avoidance, exploratory activity, and inhibition of consummatory behavior, respectively (see figure 6.6A). These assays are accepted and widely used, based on their face validity (their intuitive plausibility), their construct validity, and their sensitivity to specific pharmacological drugs. Construct validity refers to the consistency across many types of evidence that these mea-sures index anxiety states—for instance, the fact that behavior on these tests tends to be correlated, reassuring us that they are somewhat re-dundant measures of the same thing (see box 8.1). In addition, anxiety is also measured by autonomic and endocrine responses, similar to those used in studies of fear conditioning.

While scientists may know the difference between fear and anxiety, it is not completely clear whether the brain "knows" the difference. That is, scientists can formulate functional or experiential differences between fear and anxiety, but distinctions at the level of neural circuits may not be so clear cut. Nevertheless, experiments performed in the 1990s by the late Michael Davis and colleagues showed differences be-tween the effects of lesions of the bed nucleus of the stria terminalis (BNST) and the amygdala on fear versus anxiety. Using bright lights and the fear-potentiated acoustic startle response as an assay, Davis had shown that BNST lesions impaired anxiety but left fear unaffected, whereas amygdala lesions produced the converse impairment. This is a type of result called a "double dissociation," which is usually taken to demonstrate that the two processes—fear and anxiety in our case—are separate processes. Thus, the prevailing view was that the BNST promotes anxiety—persistent, tonic states, while the amygdala (in particular CeA) promotes fear—acute, phasic reactions to threats.

However, recent data indicate that the distinction between the roles of the BNST and amygdala in promoting anxiety versus fear is not as clear as was originally thought (see Tovote et al. 2015). For example, several studies have shown that functional manipulation of neurons in CeL can alter behavior in assays of anxiety (although the valence of these effects differs between reports). Therefore, increasing evidence suggests that the amygdala plays a role in anxiety as well as fear, while the BNST may be more specific for anxiety (figure 6.6 B,C). In general, whether a given brain region is claimed to be involved in "fear" or "anxiety" may depend more on the choice of behavioral assays used than on real functional differences between brain regions. Unfortunately, most papers reporting perturbation of "anxiety" circuits do not routinely include fear conditioning assays, and vice versa.

An alternative viewpoint, not excluded by the available data, is that "anxiety" and "fear" represent two different levels of intensity of a common emotion state, which are encoded by a distributed network involving interconnected regions of the amygdala and the BNST. Clearly, the circuitry and the types of behaviors that underlie fear and anxiety are complex. To disentangle them, we need to apply a battery of well-validated tests of fear versus anxiety to evaluate the consequences of manipulating neurons in different regions with cell type-specific resolution. The ingredients for achieving such a science of fear and anxiety are becoming available, but we are still a long way from having every laboratory adopt them in a uniform fashion.

A Typology of Fear States?

What do these findings imply for a functional account of fear types? What do they tell us about the kind of scheme that we saw in box 2.3,

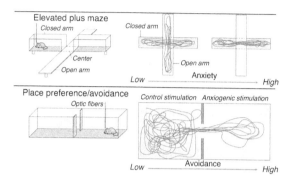

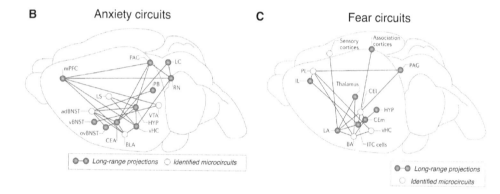

FIGURE 6.6. Behavioral assays and brain networks for fear and anxiety in rodents. (A) Three different types of behavioral assays typically used to measure anxiety. *Upper*, "open field test": anxious mice typically avoid the center of a brightly lit open space. "Optic fibers" denote fibers implanted in the brain to stimulate an anxiogenic region using optogenetics. *Middle*, "elevated plus maze": anxious mice spend more time in the closed (safer) arms of the maze. *Lower*, "real-time place preference or avoidance": mice avoid the compartment where anxiogenic stimulation is delivered. (B) Circuit nodes implicated in anxiety. Purple lines and circles denote long-range projections between nodes; yellow nodes have been further dissected into microcircuits containing specific cell types. (C) Fear circuits shown for comparison. Note the relative absence of labeling from the BNST, which in rodents is involved more in anxiety than in fear. Abbreviations (from B, clockwise): mPFC, medial prefrontal cortex; PAG, periaqueductal gray; LC, locus coeruleus; RN, reticular nucleus; PB, parabrachial nucleus; LS, lateral septal nucleus; VTA, ventral tegmental area; HYP, hypothalamus; vHC, ventral hippocampus; BLA, basolateral amygdala; CEA, central amygdala; BNST, bed nucleus of the stria terminalis (ad, antero-dorsal; v, ventral; ov, ovoid). Additional abbreviations (from C, clockwise): IL, infralimbic cortex (mPFC); PL, prelimbic cortex; CEl, lateral subdivision of CEA; CEm, medial subdivision of CEA; ITC, intercalated cell mass; BA, basal amygdala; LA, lateral amygdala. Reproduced with permission from Tovote et al. 2015.

where threat imminence was a dimension along which different types of fear could be mapped? If we find no clear neurobiological distinction between fear and anxiety, this may be telling us that there is no such functional distinction at the level of emotion states. Anxiety and fear may be more like deciding whether to turn left or turn right when running away from a predator; they are evident as distinctions in behavior (and perhaps also as distinctions in conscious experience, although this seems less clear), but they are not necessarily distinctions at the level of emotion states. Again, the fact that fear and anxiety can be distinguished operationally by scientists certainly means that there is some difference between them in the brain. But the difference may not reside at the level of emotion states, just like the difference between deciding to turn left or right is of course caused by a difference in the brain, but not by a difference in an emotion state. The upshot of these considerations is that the way we currently categorize emotions may need revision also in a direction opposite to the one we have already mentioned. We may not only need to add more fine-grained distinctions to our inventory of terms, as informed by data from neuroscience, but the data from neuroscience may also tell us that some of the distinctions that we thought there were among emotion states are not real distinctions after all.

Other Emotion States: Aggression and Anger

As we noted at the beginning of this chapter, we chose to focus on the emotion that, across a range of species, is currently the best understood in its neurobiological details: the systems that mediate fear and associated defensive behaviors. To what extent do principles revealed from those studies apply to other, less well-understood emotions, such as anger?

In humans, anger or rage are often expressed by aggressive behavior. Intermale aggression is a nearly universal behavior among sexually reproducing animal species. Is there any evidence that emotion states underlie aggression in animals? Jaak Panksepp has proposed an emotion system he calls "RAGE," that is a basic affective arousal state underlying animal aggression (Panksepp 2011a). However, there has been relatively little work to investigate whether aggression in animals is accompanied by a central emotion state. Part of the reason for this may be that, unlike studies of fear conditioning, which began in the field of psychology, studies of aggression originated almost exclusively in the field of ethology, where investigators have generally eschewed the attribution of emotion states to animals. In psychology, there are few if any papers reporting "anger conditioning" in contrast to the many thousands of papers on fear conditioning; this suggests either that attempts to create such an experimental paradigm have been unsuccessful, or that psychologists have been relatively uninterested in this topic (or both).

Is there anything we have learned about the circuitry mediating aggression that is suggestive of underlying internal states exhibiting emotion primitives? Ironically, the first experiments to demonstrate that an emotional response could be triggered by a specific brain activation in an animal involved an artificially evoked aggressive behavior. The studies were done by Walter Hess, whose work we already alluded to, and who won the Nobel Prize for his experiments. Hess electrically stimulated the hypothalamus of cats and produced a stereotyped aggressive behavior (which Hess termed "affective defense") that the cats showed to any stimulus placed in front of them (Hess and Brügger 1943).

More recent studies in mice have identified neurons in VMHvl expressing the type 1 estrogen receptor (Esr1) and progesterone receptor (PR), whose role in social behavior we already mentioned earlier in this chapter (figure 6.5). This population of hypothalamic neurons is necessary and sufficient for offensive aggression against a conspecific: silencing them prevents aggression and optogenetically stimulating them causes the mouse to attack another mouse (or, at higher levels of stimulation, an inanimate object). Neurons in this region can also promote aggression-seeking behavior in an operant conditioning paradigm, suggesting that they may play a role in encoding an internal state of aggressive motivation or arousal.

Interestingly, VMHvl Esr1$^+$/PR$^+$ neurons also play a role in male sexual behavior toward females. Although these two different social behaviors may use different, intermingled subsets of Esr1$^+$ neurons in

VMHvl, the close anatomic relationship of these populations is consistent with Tinbergen's view that offensive aggression falls taxonomically within his "reproductive" behavioral hierarchy. Indeed, aggression and mating are often seen together in animals (and for that matter, humans). Nevertheless, there is evidence that neurons in structures other than VMHvl, such as the medial preoptic area (MPOA), play a key role in male sexual behavior as well.

Positively Valenced Emotion States

All of the emotions in animals we have mentioned thus far (with the exception of states associated with mating) are aversive, negatively valenced states. But there are positively valenced emotion states as well. Much of the work on positively valenced emotion states has been performed in the context of studies of reward. The field of reward learning is a huge topic that has developed a sophisticated knowledge of multiple brain systems that control behavior aimed at obtaining rewards (defined operationally as something the animal "wants," whether or not it likes the reward after it has consumed it). It is beyond the scope of this book to cover all of this work. However, as mentioned earlier, there is increasing evidence that the basolateral and central amygdala nuclei contain distinct neuronal subpopulations that control appetitive versus defensive behaviors, and that the appetitive neurons project to brain structures implicated in reward such as the nucleus accumbens (figures 6.3B and 6.4B). It is likely that there will be significant advances over the next several years parsing the role of the amygdala in positively versus negatively valenced behaviors and associated states. Another topic relevant to this issue is hunger, which involves both motivational systems based on reducing the unpleasant sensations associated with food deprivation as well as additional motivational systems that impel the animal to obtain food because of the positive hedonic impact associated with its consumption ("liking," as opposed to "wanting" or "needing"). Each of these systems has been fleshed out in recent years using cell type-specific functional perturbations to elucidate the relevant neural circuitry, and further rapid advances in this area will be forthcoming over the next several years.

Summary

- In this chapter, we delved into some of the details emerging from current neuroscience studies of emotions in mammals. The focus was on fear, and states related to fear, and on rodents (rats and mice), since these are the systems we currently understand the best.
- We found a plethora of evidence from neurobiological studies in rodents that there were central states with properties remarkably aligned with the list of "emotion properties" we had listed in figure 3.2.
- Our review of fear conditioning led us to the brain structure that is by far the most studied for this emotion: the amygdala, which we examined in some detail. This structure is quite conserved across different species, much more so than most of the cortex, offering a good test-bed for making comparisons between rodents and humans.
- The amygdala is necessary for Pavlovian fear conditioning. The prefrontal cortex regulates emotions in part by projections to the amygdala. Fear and anxiety have historically been mapped onto the amygdala and the bed nucleus of the stria terminalis, respectively, but this distinction is likely an oversimplification and may also (or instead) reflect different aspects of a common emotion state subserved by both of these brain structures, with quantitative differences.
- There are several distinct neuronal subpopulations in the amygdala, some of which control defensive behaviors and others that control appetitive behaviors. fMRI studies in humans typically obscure this level of detail. The processing of an emotion state, while depending on specific neuronal populations, involves many such populations spread out through distributed structures in the brain; consequently, there is no single place "for fear." However, that does not mean that there are no places or neurons that play a specific role in fear.
- The hypothalamus is necessary and sufficient for innate defensive behaviors, including freezing, flight, and aggression. Specific neurons in the ventromedial hypothalamus show properties of scalability, persistence, coordination, priority over behavior control, and learning, many of the properties we listed in figure 3.2. This suggests that regions of the hypothalamus, as well as the amygdala, may encode certain emotion states.

CHAPTER 7

Emotions in Insects and Other Invertebrates

The absence of a neocortex does not appear to preclude an organism from experiencing affective states.
—*Cambridge Declaration on Consciousness, 2017 (fcmconference.org).*

Charles Darwin believed in the continuity of internal emotion states across species, all the way to invertebrates. "Even insects," he wrote, "express anger, terror, jealousy and love, by their stridulation" (Darwin 1872/1965). That belief, however, reflected shameless anthropomorphizing on Darwin's part, rather than hard scientific evidence. Is there evidence of anything like internal emotion states in insects? Have we learned anything from studies of insects that is relevant to our understanding of emotion states in other animals, including humans? The answers to these questions are not yet clear. However, if for no other reason than the fact that Darwin was usually right in his other intuitions about biology and evolution, it is worth considering what has been learned from modern studies of defensive behaviors and associated internal states in insects.

There are at least two reasons why studying emotions in invertebrates is so interesting and important. One reason is that it forces us to incorporate an abstract level of analysis into our account of emotion. Ironically, even though Darwin was tempted to anthropomorphize the emotional behaviors of insects, taking another look at these animals in the framework of this book forces us to adopt a broader and functional view that needs to understand a nonhuman ecology. Invertebrates like insects do not have a neocortex (only mammals have a proper neocortex). In fact, invertebrates do not even have an amygdala, nor a hypothalamus; their brains are truly alien compared to ours. If we ever want to build a robot with emotions, we had better understand how emotions work in an insect brain.

A second reason for the interest lies in our need to understand how emotions arose in evolution, and which building blocks were the first to appear. Very simple nervous systems, such as the net-like architectures seen in jellyfish and the small nervous systems of worms, probably arose 540 to 600 million years ago. Some of these animals are subjects of intense study in modern neuroscience. The worm *C. elegans*, for instance, has had all the connections between the roughly three hundred neurons of its nervous system mapped in detail. Remarkably, the ion channels, neurotransmitters, peptides and other molecules expressed by *C. elegans* neurons are quite similar to those expressed in mammalian neurons—with the notable exception of voltage-gated sodium channels, which are lacking in the worm.

While studying species very different from us is thus informative for several reasons, it is important to keep in mind that comparative neurobiology is not the same as having a time machine. We can study all the diversity of animals that have nervous systems, but they are all modern, derived species—not the ancestral species from which this diversity evolved. While this comparative approach helps us to abstract broad functional features, it is important to bear in mind that the brains of other animals—especially invertebrates—are not "less evolved" than our own. Rather, they have evolved to perform functions that are best adapted to those particular species' survival and reproduction. The nervous systems of contemporary insects are as "modern" as the nervous systems of humans.

Learned Avoidance Behavior in *Drosophila*

The fruit fly (or more accurately, vinegar fly, since the animal is attracted by the smell of vinegar produced by ripe or rotting fruit), *Drosophila melanogaster*, has been a key model organism in the study of mechanisms of heredity and development, garnering several Nobel Prizes for investigators who have used its experimentally tractable genetics to investigate those mechanisms. In the 1960s, Seymour Benzer, a molecular biologist at Caltech who made fundamental contributions to our understanding of the structure of genes, began to establish *Drosophila* as a system for studying how genes control animal behavior. Those studies established an entire field that now numbers in the thousands of investigators, and which was recognized by awarding the 2017 Nobel Prize in Physiology or Medicine to Michael Rosbash, Jeff Hall, and Michael Young for their identification of genes that control circadian rhythms, starting with mutant flies originally isolated by Benzer and his student Ron Konopka in the early 1970s.

Among his many other contributions, Benzer and his students showed that *Drosophila* were capable of aversive associative learning; flies that are administered an electric shock during exposure to a particular odor will learn to avoid that odor. This observation was surprising to many people at the time, who had assumed that flies were little automata incapable of anything other than reflexive behaviors. Benzer and his colleagues, moreover, went on to show that this associative learning was dependent on some of the same molecules implicated in associative learning in other organisms (for example, the sea snail *Aplysia californica*, studied by Nobelist Eric Kandel and his colleagues). These pioneering studies, therefore, demonstrated an evolutionary conservation of Pavlovian aversive conditioning as well as a conservation of some of the underlying molecular mechanisms. But what have we learned since that time about the neural mechanisms that underlie associative learning in flies? Have those studies taught us anything of relevance to emotional learning in higher organisms?

Olfactory avoidance learning in *Drosophila* is superficially similar to fear conditioning in rodents: an odor is the CS, and a shock is the US. (Odorants can also be used as CS's for appetitive conditioning in *Drosophila*, with sucrose as the US.) The powerful genetic tools available in this organism, combined with new technologies for manipulating neuronal activity such as optogenetics (box 5.1), have begun to provide important insights into the organization of neural circuits underlying these learned behaviors. As is the case in the rodent fear-conditioning literature, studies of olfactory conditioning in *Drosophila* thus far have taught us arguably more about learning and memory than about emotion. Nevertheless, important findings have emerged that have illustrated mechanisms by which brains can attach a valence to a previously neutral stimulus.

A fly can be trained to approach or avoid a neutral odor, depending on whether the US paired with the odor is appetitive (sucrose) or aversive (shock). How can the fly's brain attribute either a positive or negative value (valence) to the same stimulus, depending on its experience? Early work in flies by Martin Heisenberg provided evidence that the neurotransmitter dopamine (DA) was involved in aversive conditioning, while octopamine (the insect equivalent of the neurotransmitter norepinephrine found in vertebrate animals) was involved in reward learning. By contrast, in mammalian studies, DA was thought to be involved exclusively in reward learning. This finding was interpreted by some to mean that the DA neurotransmitter system might encode opposite valences, in vertebrates versus invertebrates. More recent studies, however, have shown that DA neurons control *both* appetitive *and* aversive conditioning in flies (although octopamine is also involved in the former). This is possible because genetically and anatomically distinct subpopulations of DA neurons are required for the two different forms of learning (figure 7.1A1 versus B1, DANs) (Owald and Waddell 2015). Indeed, a similar story has emerged from studies in rodents and primates: whereas DA was originally linked to motivating behaviors to obtain rewards (seeking and learning) it is now clear that DA also participates in arousal and salience (orienting and attention), and even in rapid detection of aversive stimuli. Once again, these different facets of DA's role in emotions appear to be mediated by distinct neuronal subpopulations (Bromberg-Martin, Matsumoto, and Hikosaka 2011).

While much remains to be learned about the detailed mechanisms of plasticity in the fly DA system, there are several important general conclusions that can be drawn. First, and most important, the emotional valence of a given US is not encoded merely by the identity of a particular neurotransmitter; DA does not intrinsically encode reward or pleasure, as commonly believed. Rather, the intrinsic valence of a given US (shock or sucrose) is determined by *which* DA neurons it activates, and whether these DA neurons influence synapses with neurons that mediate approach or avoidance (Aso et al. 2014); this connectivity is likely genetically specified and emerges during development as the brain is wired together. That is, it is not the identity of a chemical in the brain (for example, dopamine), but the connectivity of the neurons that release that chemical, which determines behavior. It is, ultimately, the function of neural circuits that specify what is being contributed to an emotion state.

Second, in contrast to a prevailing view, DA does not act as a "sprinkler system" in the brain, acting via so-called volume transmission that simply douses a swath of neurons in a diffuse manner. Rather, it acts in a highly spatially compartmentalized manner at specific synapses, which are determined by the genetic identity and wiring of a particular class of DA neurons—a mechanism of action more often associated with classical neurotransmitters such as glutamate, than with neuromodulators like dopamine.

Third, the brain's ability to associate a given odor (CS) with a US of either a positive or negative valence depends upon (a) the *sparse representation* of the odor by Kenyon cells (that is, the activity of a relatively small number of highly odor-specific neurons), and (b) the innervation by a given CS-responsive Kenyon cell of multiple *compartments*, where they make synapses with mushroom body output neurons (MBONs) that promote *either* approach or avoidance, in a compartment-specific manner (figure 7.1, $\gamma2$, $\gamma3$, $\gamma4$, $\gamma5$). As shown by elegant work from Vanessa Ruta and colleagues, these Kenyon cell–MBON synapses can be either selectively strengthened or weakened, according to which DA neurons are activated by the US and in which compartment(s) the DA is released (figure 7.1, A2 versus B2). Interestingly, evidence in mice indicates that different subpopulations of DA neurons within the ventral tegmental area, traditionally implicated exclusively in reward, can control either reward or punishment (but not both) (Lammel et al. 2012). So, one general principle that emerges from *Drosophila* is that molecules (such as DA) do not encode valence; rather, it is neuronal connectivity that determines whether a particular stimulus is rewarding or punishing.

The studies described above have provided important insights into how the brain can flexibly attach either a positive or negative valence to a previously neutral stimulus, at a level of mechanistic detail that has not yet been achieved in studies of any mammalian system. This adaptive function is fundamental to an animal's ability to learn whether to approach or avoid an initially unfamiliar object or stimulus. However, in order for such a system to function, the brain has to be "hardwired," at least to some degree, with innate representations of positive and negative valence—in this case the different classes of DA neurons

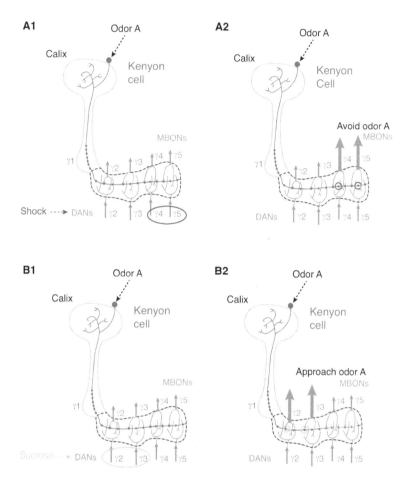

FIGURE 7.1. Compartmental release of dopamine controls appetitive versus aversive conditioning. A given odor (odor A) activates a mushroom body (MB) Kenyon cell, whose axon synapses with MB Output Neurons (MBONs) in four different compartments of the gamma lobe (blue line/blue dots). These four different compartments of the gamma lobe are labeled γ 2–5 in the figure. The MBONs are hardwired to evoke approach (γ2, γ3) or avoidance (γ4, γ5). The activation of the Kenyon cell by Odor A, in the absence of a US (shock or sucrose), is too weak to evoke activity in the MBONs. However, pairing Odor A with a US causes local release of dopamine (DA) in specific compartments, which facilitates synaptic transmission between the Kenyon cells and MBONs in that compartment. Odor A can be paired with either a punishing US (shock, A1) or a rewarding US (sucrose, B1). The punishing US activates a different subset of DA neurons (DANs; γ4 and γ5) than the rewarding US (DANs γ2 and γ3). Because these DANs innervate different compartments of the gamma lobe, their activation selectively modifies KC → MBON synapses in the corresponding compartments (red vs. blue ovals in A1 vs. B1, respectively). Following training with a shock (A1), presentation of odor A preferentially activates MBONs innervating the γ4,5 compartments, leading to avoidance (A2). Conversely, following training with sucrose (B1), presentation of odor A preferentially activates MBONs innervating the γ2,3 compartments, leading to approach. Modified with simplifications from Cohn, Morantte, and Ruta 2015.

that mediate reward versus punishment—which can then be linked to stimuli whose valence must be learned by experience (Aso et al. 2014). Without some such innate basis, there would be nothing to ground valence, nothing upon which learned associations could build. Those innate representations of valence, in turn, would not have been selected in evolution if they did not afford the species a survival advantage. Natural selection acted to link sensory circuits that detect specific, ecologically relevant stimuli, to motor circuits that trigger appropriate responses (approach or avoidance) through the activation of such valence representations. Whether those same valence systems are involved in learned avoidance or approach remains to be investigated.

Does *Drosophila* Have Emotion States?

But does the activation of a neural system that represents valence equate to the activation of an internal emotion state? Does the fact that a fly invariably escapes from a looming object, on a timescale of tens to hundreds of milliseconds (Card and Dickinson 2008), necessarily mean that the stimulus has triggered an emotion state? We would argue that it does not, unless some other properties in addition to a representation of valence are also present (chapter 3). How many other properties are needed? Here, we would acknowledge that in simpler animals like flies there may be simple emotion states, composed of "emotion primitives," that amount to a few of the properties we listed in figure 3.2—less than the full set, but more than just one of them. These primitives, which one could think of as "building blocks" of an emotion state, might feature, in addition to valence, also persistence, scalability, integration, or generalizability (the flexible influence of the state on behavior in multiple contexts). Is there any evidence of such emotion primitives in the behavioral responses of flies to innately aversive stimuli?

A recent study investigated this question by quantitatively analyzing the behavioral responses of flies to multiple presentations of an overhead dark translational (sweeping) stimulus (Gibson et al. 2015). Such stimuli are remarkably effective in triggering innate escape or avoidance responses, not only in flies but also in mice and other mammals, presumably because they activate visual circuits that evolved to detect

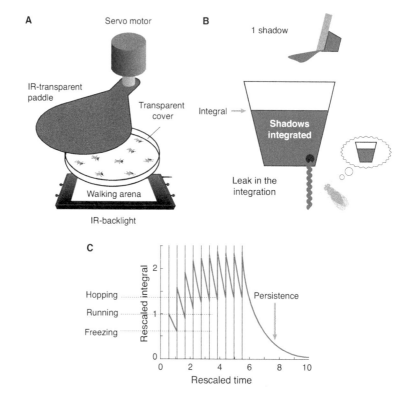

FIGURE 7.2. Experimental system for investigating whether innate defensive responses in flies display any emotion primitives. (A) Flies in a covered arena are exposed to repeated presentations of a translationally moving dark overhead object. They respond to increasing numbers of stimulus presentations with freezing, increased locomotor activity and hopping, but cannot escape. Following offset of the stimulus, the defensive behavior of the flies persists for some time before settling down. (B, C) The behavioral responses of the flies to the stimulus can be modeled by assuming that the fly's brain behaves as a "leaky integrator" of stimulus exposure. (B) Analogy to a leaky bucket; each time the stimulus passes overhead, more water is poured into the bucket, which leaks at a fixed rate. If the rate and frequency of stimulus exposure exceeds the leak rate, the bucket will fill, producing maximum levels of defensive behavior. Once the stimulus ceases, defensive behavior will persist until the bucket drains. This model has similarities to the hydraulic model proposed by Konrad Lorenz to explain internal "drive states" (see figure 5.2), except that the latter posits that the 'bucket' is filled automatically, rather than in response to external stimuli, and no persistence is invoked. This model can explain how the accumulation of information about threat results in specific behaviors. (C) Mathematical formalization of the model. The y-axis represents the value of the integral (amount of water in the bucket in [B]), the x-axis represents time. Each vertical line represents one pass of the overhead stimulus. The dashed lines illustrate how different defensive behaviors have different thresholds, which are successively crossed by the system with accumulation of information, progressing from freezing to faster walking to hopping. The model accounts for the emotion properties of scalability and persistence. The response to the shadow also displays a negative valence, and generalizes to different situations. Reproduced with permission from Gibson et al. 2015.

the approach of an aerial predator (remember figure 2.5). However, it is difficult to search for properties like persistence and scalability following single presentations of such stimuli, since the flies escape and cannot be recovered. Therefore, in order to examine the flies' response to multiple presentations of a threatening stimulus, they were enclosed within a large plexiglass arena with a transparent cover (figure 7.2A).

Strikingly, flies became progressively more "agitated" (active), measured as an increase in their locomotor activity, as the number of exposures to the threatening visual stimulus (that is, paddle sweeps) was increased. Moreover, the nature of the behavioral response changed qualitatively with the number of stimulus exposures: a few passes of the overhead shadow caused the flies to freeze or increase their walking speed, but with further successive exposures the flies switched to hopping. Furthermore, after the last of a series of stimuli were delivered, the flies' response *persisted* for tens of seconds to minutes; the insects continued to hop and walk at high velocity, and then gradually "calmed down" over a period of many tens of seconds. These observations provide evidence of *scalability* and *persistence* in the flies' response to the threat stimulus.

Interestingly, this cumulative effect of multiple stimulus exposures was dependent on the interval between stimulus delivery (shadow passes): if the interval was too long (> 1s), no cumulative effect of the stimulus was observed. This dependence on interstimulus interval, and the gradual decay of the response (persistence), is suggestive of the operation of an underlying "leaky neural integrator" of visual threat exposure, somewhere in the fly's brain. An analogy is to a leaky bucket that is periodically partially filled with water from some intermittent source; if the amount of water added each time, and/or the frequency of fills, exceed the leak rate, then the bucket will gradually fill even though it is leaky (figure 7.2B). Once the filling has stopped, the bucket will slowly drain. A mathematical model based on this concept produces similar scalability and persistence as observed in living flies exposed to repetitive visual threats (figure 7.2C). It is similar to the drift-diffusion models of decision-making that we introduced in chapter 3 (figure 3.4).

The negative valence of the stimulus was established by exposing flies to the visual threat while they were on a central food patch, feeding.

Exposure to multiple threat stimuli eventually dispersed the flies from the food patch, in a direction away from the leading edge of the stimulus. Interestingly, flies exposed to ten presentations of the threat stimulus took longer to "calm down" and return to the food patch than did flies exposed to only four presentations of the stimulus, suggesting that ten stimulus presentations may have caused a higher level of "fear" than did four presentations. These observations are suggestive of a decaying internal state of defensive arousal or threat alert that interferes with feeding, whose initial magnitude is proportional to the integrated strength of the stimulus, in the "leaky bucket" model (figure 7.2B).

Taken together, these data suggest that the behavior of flies exposed to multiple successive threat stimuli is not purely reflexive, but rather exhibits multiple emotion primitives (chapter 3), including valence (negative in this case), scalability, persistence, influence on decision-making (whether to remain on or leave a food patch), automaticity, priority over behavioral control (inhibition of feeding or mating), and ability to generalize to different contexts (with and without food). This is consistent with (but does not prove) the operation of an underlying internal state with similar properties, that causes these behaviors. One would next like to find independent evidence for such an underlying internal state, a clear direction for neurobiological measures in future experiments. At the very least, these experiments demonstrate a way to experimentally test for some of the behavioral features that display properties characteristic of emotional expression, which distinguish such expression from simple reflex behaviors, even in a fly. They also provide quantitative behavioral data that support Darwin's intuition that even insects have emotion-like states (Darwin 1872/1965).

What adaptive value might such a scalable, persistent, and negatively valenced internal state have? Consider hungry flies sitting on a piece of ripe fruit: a shadow passes overhead and the flies have to "decide" whether to continue feeding or to escape. This is not a trivial decision: escape requires energy, negating the caloric benefit of the food consumed; therefore, premature escape from a "false" stimulus that does not actually signal a predator could eventually cause death by starvation and would be maladaptive (see box 2.2). On the other hand, if the flies remain on the food for too long, until multiple experiences of the stimulus have increased the certainty that a real threat is indeed present, they risk death by predation. Thus the choice of whether to stay or flee literally amounts to a life-and-death decision and requires that the fly's brain perform a kind of cost-benefit analysis to decide whether to continue to feed or to escape, each time they encounter a threat. The ability to integrate multiple stimulus exposures over time and translate that into a state variable whose intensity determines whether to stay or flee is likely to be critically important to this cost-benefit analysis.

What do we know about the neural circuitry that underlies the behavioral responses of the flies under these conditions? Do flies have the equivalent of an amygdala or a hypothalamus in their tiny brain that integrates the influences of diverse types of aversive stimuli? We do not yet know. The fly brain (with some notable exceptions) is organized very differently from the mammalian brain, and generally does not have central structures that correspond directly with (that is, are homologous to, in the evolutionary sense) those in our brains. Nevertheless, there are analogous areas; for example, the mushroom body may be analogous to the hippocampus and/or olfactory cortex, the antennal lobe is similar to the mammalian olfactory bulb, and the retinae of flies and mammals display remarkable organizational similarities even though they evolved independently. Furthermore, the fly brain is made up of neurons that are, from the molecular standpoint, very similar to those in our brain. So, while the fly brain may not look superficially like a mammalian brain, it has analogous functional areas, uses the same chemicals (dopamine, serotonin, acetylcholine, GABA), and its neurons are made of the same molecules as mammalian neurons. Therefore, the question is not "does the fly have an amygdala?" The question is whether there is a common circuit node that processes defensive responses to many different kinds of threatening stimuli, or whether each type of stimulus activates its own "private" response pathway.

One well-known example of such a "private" pathway is that controlling escape from a looming visual stimulus. *Drosophila* have a well-described circuit that mediates a rapid, reflexive jump away from a looming visual threat, consisting of a large "descending interneuron," called the "giant fiber," that extends from the central brain to the thoracic ganglia (the fly's functional equivalent of the spinal cord). Information travels quickly through electrical synapses (gap junctions) along this pathway from the fly's visual system to the motor neurons that activate the jump muscles, allowing an escape response to occur within a few tens of milliseconds of detection of the threat (this is what makes flies so hard to swat). As far as we know, this pathway is specific for threatening visual stimuli and is not, for example, activated by aversive odors. However there is now evidence of multiple circuits that mediate escape responses to visual threats in flies (Reyn et al. 2014). Perhaps some of those circuits also process escape responses to stimuli of different sensory modalities. This question is likely to be answered in the near term. Another important question for the future will be to understand how the circuits that mediate reflexive defense responses are related to those mediating integrative, emotion-like responses: are these independent, parallel pathways, or do they reflect state-dependent modification of the same circuit?

Anxiety in Insects and Other Arthropods

As in the case of mice, there is a distinction between the way that "fear" and "anxiety" have been studied in invertebrate systems. Flies show thigmotaxis; when placed in a large (relative to the size of a fly), open arena, they tend to walk hugging its walls. This behavior appears superficially similar to the avoidance of the central region of an arena exhibited by "anxious" rodents in the so-called open-field test. Administration to rodents of drugs that reduce anxiety in humans, such as Valium (a benzodiazepine that targets GABA receptors), makes the animals spend more time in the center of the arena (see figure 5.3 and chapter 6). Remarkably, thigmotaxis in flies is also reduced by treatment with similar anxiolytic drugs targeting GABA or serotonin receptors (Aryal et al. 2016). These results were interpreted to provide "pharmacological validity" that the thigmotactic behavior expressed "anxiety" in flies. However, not all instances of thigmotaxis necessarily indicate anxiety; for example, the wall-following behavior could reflect the self-perpetuating effects of optic flow (caused by movement of the wall along the fly's retina) to promote locomotor activity along the perimeter of an arena, via an "optomotor response," rather than result from anxiety-based avoidance of the open central region. In that case the "anxiolytic" drugs (which act on receptors distributed widely across the brain) might interfere with visual circuits mediating this optomotor behavior.

Another study investigated crayfish, which like flies are arthropods, using an aquatic version of the elevated plus maze to assess their behavior. (In an elevated plus maze, one pair of arms—for example, the N–S axis—is open and brightly lit to promote aversion while the perpendicular pair of arms—for example, the E–W axis—is enclosed and safe.) It was reported that exposure of the crayfish to a benzodiazepine anxiolytic drug reduced avoidance of the illuminated arms of the maze, while injection of serotonin or exposure to inescapable shock increased avoidance of those arms (Fossat et al. 2014). This response is similar to the effect of these manipulations on mice or rats tested in a terrestrial version of the elevated plus maze (see figure 6.6). While intriguing, these studies continue in the vein of trying to search for states that are homologous to specific emotion states in humans by applying analogous behavioral or pharmacological criteria, rather than by searching for species-typical behaviors that display general properties characteristic of emotional expression.

Emotion States and Social Behavior in Insects

Mating and aggressive behavior in mammals are associated with high-intensity internal states of motivation, arousal, and drive (see chapter 5), and in humans with subjective feelings we label these states with words such as *love, lust, anger,* or *rage.* What are the neural systems that operate to control reproductive behaviors in *Drosophila,* and if so do they play a role in the instantiation of internal states? One basic question is whether flies engage in sex because it is rewarding or reinforcing to them, or simply because they are genetically programmed to do it. An interesting study by Ulrike Heberlein and colleagues showed that male flies were preferentially attracted to an odor they experienced during mating with a female, compared to a control odor. Similar results were obtained when the odor was paired with ethanol, which flies prefer to drink. These results were interpreted to suggest that both mating and ethanol consumption are rewarding to flies. However it is possible that the US for associative conditioning was the smell of a virgin female, not the experience of mating per se.

Male flies exhibit aggressive behavior toward other males, as well as courtship and mating toward females. Recent studies have identified an interconnected network of sexually dimorphic neuronal clusters that control male courtship behavior in this species. (By "sexually dimorphic," in this case, it means that the neurons are present in males but not detectable in females.) One of these clusters, called P1, not only promotes male courtship behavior but also can promote aggression. Importantly, optogenetic stimulation of P1 neurons can trigger a persistent internal state in solitary flies, which promotes aggression once the fly encounters a conspecific male. This internal state can endure for tens of minutes in the absence of social contact and may represent a type of persistent memory of a mating encounter with a female, which can trigger aggression when a competing male is encountered. These data suggest that courtship and aggression in *Drosophila* are not simply reflexes, but are associated with persistent, internal states (Anderson 2016). Given the finding that mating and aggression are also controlled by spatially colocalized and anatomically intermingled neurons in mammals (chapter 6), these results also suggest that the neural circuitry underlying these innate social behaviors may exhibit similar organizational features in flies and mice, species separated by 500 million years of evolution. Whether this similarity reflects divergent (descent from a common ancestor) or convergent evolution remains to be determined.

Internal States in Other Invertebrates

An alternative approach to searching for evidence of emotion states in insects is provided by studies of "cognitive bias" in bees (Mendl, Burman, and Paul 2010) (see box 3.3). Here the idea is that emotions can be identified as internal states that alter "cognitive" behavior, such as decision-making under conditions where the outcome of the decision is ambiguous. For example, it has been shown that disturbing bees by violent shaking causes them to choose a less favorable outcome (that is, to select a cue less likely to predict a reward). This has been interpreted as evidence that the shaking caused a "pessimistic" cognitive bias, in other words, a negatively valenced emotion that made the bee behave as if it was unlikely to be rewarded (Bateson et al. 2011). Similar experiments have suggested that, conversely, high concentrations of sucrose can promote a "positive" cognitive bias in bees. It is clear that bees have impressive cognitive abilities and sophisticated behaviors (their brains contain close to a million neurons). However, it can be difficult to distinguish the effect of a relatively "blunt instrument" stimulus, such as shaking or sugar, to promote an emotion state that biases a higher-order cognitive process, from one that alters peripheral sensory acuity or motor activity in a manner that influences performance of the behavior. Among invertebrates, the only other organism that has genetic tools comparable to those available in *Drosophila* is the nematode (round worm) *Caenorhabditis elegans*. This animal, which was introduced as a model organism by Nobel laureate Sydney Brenner in the 1960s and '70s, has the advantage that it is very small (~1 mm long), transparent, has a short generation time (three days) and only 302 neurons composing its entire nervous system. Moreover, the complete "connectome" (neuronal wiring diagram) of the worm has been determined by reconstruction of serial electron micrographs; it is thus far the only adult organism for which a complete connectome has been established.

Whether or not *C. elegans* has emotion states is not yet clear, but it certainly does have persistent internal states that influence its behavior, which have been analyzed at a level of molecular and cellular detail not yet achieved in any other organism. Work from the laboratory of Cornelia Bargmann, for example, has shown that the organism alternates between two extended states, called "roaming" and "dwelling" (Flavell et al. 2013). As their names imply, these states are engaged when a worm is foraging for food, or is relatively sedentary. Serotonin (which is well known to influence mood and arousal in mammals) promotes the dwelling state by inhibiting neurons that promote roaming. Conversely, a neuropeptide called pigment dispersing factor (PDF) promotes the roaming state. Optogenetic activation of the intracellular signaling pathway engaged by PDF can trigger roaming behavior lasting minutes in animals that were dwelling prior to stimulation. These studies show how specific neuromodulators, including a biogenic amine and a neuropeptide, act on neural circuits to promote persistent changes in behavioral state and begin to reveal some of the most fundamental aspects of how emotion primitives can be encoded in nervous systems.

As illustrated in the foregoing examples, invertebrate organisms can be used to study the neurobiology of emotion primitives, rather than to mimic or model specific human emotions. Our coverage of this topic has been superficial, and there is a great deal that other invertebrates, such as cephalopods, can contribute to the study of emotion (box 7.1). The strength of the systems described here reflects the fact that we can use experimental stimuli and dependent measures unavailable in humans, and that we can use neurobiological tools, like optogenetics, calcium imaging, electrophysiology, and automated analysis of behavior, that permit a level of cellular resolution and causal inference that cannot be achieved in humans. To the extent that there is any evolutionary conservation or universality of specific emotion states, it seems wisest to focus, for the time being, on those states that offer the clearest criteria for their assessment, and an obvious adaptive value to the survival of the species. "Fear," or a persistent internal state of defensive arousal or threat alert, is a good example of such an emotion; awe or shame would not be. When we turn to human studies in the next two chapters, we will see somewhat of an inverse of this emphasis. The studies in humans depend on methods and measures that are in many ways inferior to those that can be applied to animals, but the array of emotions studied, and ways of inducing them, is extremely rich. Animal studies (as reviewed in this and the previous chapter) tend to offer very detailed cellular-level accounts of mechanisms for a specific aspect of only a very few, specific emotion states. By contrast, human studies tend to offer much broader and less mechanistic, systems-level accounts of emotion, but prominently include richer aspects that are important to our lives, like emotion regulation and feelings. As we have noted repeatedly, relating these two very different approaches to the study of emotion with one another is one of the most important challenges for the future of a science of emotion.

BOX 7.1. Internal states of cephalopods.

Cephalopods, including octopuses, squid, and cuttlefish, have a complex nervous system that looks nothing like a vertebrate nervous system, and they exhibit complex behaviors. They are about as close as we can currently get to intelligent aliens. An octopus may have about 500 million neurons in its brain, more than a mouse (ca. 100 million) and far more than a fly (ca. 250,000), but far fewer than a human (about 80 billion).

Octopuses in particular have been reported to be capable of sophisticated behaviors, including observational learning and problem solving. It is tempting to speculate that the rapid changes in pigmentation that can be observed when octopuses are threatened in captivity reflect some type of emotional expression, but there is evidence that these changes may be controlled peripherally, in the skin, rather than by the brain. It is thus possible that, while these animals show functional criteria for emotions, they do not have a centralized, internal neurobiological state as vertebrates do. As with other aspects of their behavior, control may be much more distributed—giving the scientist an interesting challenge in how to understand such a system.

Both octopuses and squid exhibit a variety of defensive behaviors, including crypsis (camouflage), ink jetting, and propulsive escape. The context in which these different behaviors occurs can depend on the proximity of the animal to the threat, as documented by Roger Hanlon at the Marine Biological Laboratory in Woods Hole, Massachusetts. Crypsis occurs when a threat is distant, but switches to escape (ink jetting and propulsion) as the threat becomes more proximal. These changes may correspond to the switch from freezing to flight behavior that is characteristic of terrestrial animals and fit the same threat imminence theory that we described in box 2.3. These transitions in defensive behavior as a predator or threat becomes more proximal could represent a form of scalability, in that an internal state of defensive arousal would presumably increase as the predator approaches.

Whether octopuses exhibit other emotion primitives, including persistence, valence, and generalizability, has not yet been studied systematically. One of the challenges in studying the neurobiology of emotion in octopuses is that these creatures are difficult to maintain and breed in the laboratory (they have a remarkably short life, only a year or two), are not yet amenable to genetic manipulations, and are difficult to adapt to preparations for electrophysiological recording or imaging. Nevertheless, they are fascinating creatures, and one looks forward to a time when the neurobiology of emotion and internal states in these elusive organisms becomes a mature science.

Summary

- *Drosophila melanogaster*, the fruit fly, can be used to elucidate the neural circuitry and chemistry underlying aversive associative learning, a behavior similar in its features to Pavlovian fear conditioning in mammals.
- Innate defensive responses to visual threats in the fly consist of rapid reflexive reactions as well as of integrative responses that exhibit several emotion building blocks also seen in mammalian species, including scalability, persistence, valence, generalizability, and priority over ongoing behavior. These building blocks may indicate an underlying internal emotion state.
- The neurobiology of a persistent internal state can be dissected in elegant mechanistic detail using *C. elegans*, a genetically tractable nematode worm containing only 302 neurons in its entire nervous system. While worms and flies are unlikely to be useful for studying social emotions like shame, these studies illustrate how "model organisms" in biology offer a trade-off between experimental tractability and the complexity of the phenomena that can be investigated.

CHAPTER 8

Tools and Methods in Human Neuroscience

Just like the vast majority of emotion studies in animals has historically come from behavioral studies, so too have the vast majority of emotion studies in humans. Yet these two historical lines of work have mostly used very different kinds of measures: in animals, standardized behavioral tests in the laboratory, or ethological observations in natural environments, and in humans, studies of self-reported emotion in the laboratory. This has made it difficult to connect studies of emotion in animals with those in humans.

Given what you just read about in the previous four chapters, one might hope that a neurobiological approach to emotion would bring the study of animals and humans closer together. Don't all neuroscientists use the same tools? Well, even here there are large differences, as it turns out. We survey some of the approaches used to study the brain in human neuroscience studies of emotion in this chapter and provide comparisons with the approaches used to study the brain in animals that were the topics of chapters 4–7. A major challenge will be how to bridge data from tools like functional MRI (mostly used in humans), which measures correlates of blood flow in large regions of the brain, and electrophysiology, calcium imaging, or optogenetics (often used in animals), which measure activity in single neurons and manipulate specific neuronal cell populations (see figure 4.2).

One way to relate the different kinds of neuroscience methods to one another is by rephrasing our questions at a more abstract level. In chapter 3 we discussed the processing features that central emotion states should possess in order to carry out their functional roles, and these properties of emotions outlined abstract computational or architectural constraints that we could begin to look for in brains. In chapter 4, we discussed how such a more abstract computational or algorithmic level of analysis could be related to neuroscience measures (figure 4.4). In chapters 5, 6, and 7 we saw how, indeed, there are specific circuits that cause emotional behaviors in animals, and that these circuits do show many of the properties we listed for emotion states in figure 3.2.

The hope is that similar descriptions that bridge different levels of analysis could be achieved also for the human brain. We want a story that links functional criteria (what is the adaptive role that an emotion was selected for?) to processing features (what are the computations required to fill that functional role?) to implementation discoveries (how is this actually done in the brain?). Very roughly, these three levels of abstraction correspond to something like the level that the ethologist uses when observing behavior, the level that the cognitive- or systems neuroscientist uses when discussing processing features and architectures, and the level that the neurobiologist uses when collecting data from the brain. However, most systems and cognitive neuroscientists (the authors of this book included), in fact use analyses across all three levels.

There is, of course, one huge apparent advantage in human studies of emotion, compared to studies in animals: humans can speak and understand language. This fact, together with other practical and ethical considerations, introduces great differences in methods into neuroscience studies of emotion as done in humans as compared to animals. Human subjects are typically given brief verbal instructions on what to do, often indicate what they are doing (or that they have understood what to do) by verbal discussion, generally don't put up with days of training, cannot easily be presented with stimuli that evoke intense negative emotions for ethical reasons, and often evidence an emotion state through verbal reports on how they feel. So there is little or no training, the emotions are typically very faint compared to what is studied in animals, and language processing is usually involved to a very considerable extent.

This makes studies in humans often much easier to carry out than in animals: you just tell your subject what to do rather than arranging an elaborate situation or training paradigm ("think of a time in your life you felt really sad, and tell me how you feel"). However, this approach also easily seduces the experimenter into believing that subjects are indeed doing what the experimenter thinks they are doing, which is not always the case. For example, it is critical to carefully interpret what subjects are saying, since verbal reports of their experiences are data that are often highly susceptible to the subject's background beliefs and expectations, including the expectations introduced by the experiment. For instance, if you are being paid money as a research subject and told to feel sad, you might well report feeling sad because you want to earn the money, or because of the general social expectations of the experiment (so-called demand characteristics). Psychologists are well aware of these complications and have devised questionnaires and other dependent measures that can help triangulate on an emotion state. For these measures to convincingly assay an emotion state, such as fear, they have to show strong *construct validity*—typically highly correlated results across a range of different dependent measures, not solely verbal report (box 8.1).

BOX 8.1 Construct validity.

Scientists use particular measures to infer the existence of certain variables in their theories. For instance, physicists use a lot of fancy instrumentation to infer the nature of subatomic particles. Psychologists and biologists might use behavioral or brain measures to infer the existence of constructs like attention or memory—or emotion. This raises an obvious question: how do you know your measures are good indicators of what you think they measure?

The ability of a set of measures to yield good evidence about a certain construct is called "construct validity." This is actually a bundle of criteria. One important aspect of validity is "face validity," which is simply the plausibility of any specific measure to provide evidence for a particular construct. If I have a detailed questionnaire about your feelings of fear and anxiety, this has a reasonable face validity as a measure of fear. If I measure fMRI signal in your amygdala, this has questionable face validity as a measure of fear without a lot more information. If I only measure your blood pressure, this has poor face validity for fear (by itself, since we can easily think of cases where blood pressure has no relation to fear).

These examples highlight the need for multiple good measures and lead to the main two aspects of construct validity: convergent validity and discriminative validity. If we want to have convincing evidence for a state of fear, only measuring blood pressure or facial expression is not a very reliable indicator. However, if we measure blood pressure change, heart rate change, ratings on a fear questionnaire, facial expression, and fMRI signal, we might be able to use all of these together to get quite convincing evidence for fear. If these measures all turn out to be correlated during a fear state, this would provide convergent validity; whatever they are measuring, they seem to be measuring the same thing, and if at least one of them has face validity for measuring fear, then so do the others when taken together. On the other hand, we would also want these measures to take different values for a different emotion—the blood pressure, fMRI signal, heart rate, facial expression, and self-report should look different if the person is feeling, say, happiness rather than fear. Many of these measures, in isolation, may have very poor discriminative validity. For instance, the typical autonomic measures collected in the lab (heart rate, skin conductance, etc.) do not clearly distinguish between different emotions (cf. box 2.1), although there is active research on whether they might provide discriminative validity if we look at patterns across multiple measures.

These issues of construct validity are particularly important in human neuroimaging studies, because there is often no overt behavioral measure, unlike in animal studies. Since the subject must lie immobile in the brain scanner, people cannot be imaged if they are screaming, fighting, or running away. This makes it especially important to ensure that other dependent measures (measures of autonomic activity, self-report, facial expression) are well validated, and ideally are all collected together and provide consistent evidence for the emotion that the experimenter intends in the study.

Historical Neuroscience Studies of Emotion in Humans

Before the advent of modern cognitive neuroscience (in the 1970s and '80s), and very much before the advent of fMRI (in the mid-1990s), there was a smattering of case reports that appeared to provide support for the hypothesis that there are specific anatomical regions in the human brain subserving emotions. Earliest among these were rare neurological cases—in the clearest cases, patients who had lesions to specific parts of the brain (as was also the case for other domains of cognition, like language and vision and memory).

Probably the most famous lesion case with relevance to emotion was Phineas Gage, who has been described in books and articles and whose brain lesion has been reconstructed from the little evidence we have available (Damasio et al. 1994; MacMillan 2000). Briefly, here is Gage's story. Gage was a foreman working on building a railroad line in Vermont, for which tracks had to be blasted through the rock. To do this, a hole was drilled into the rock, partly filled with gunpowder, tamped down with an iron rod, and then exploded from a safe distance. On a fateful day in 1848, a rather gruesome accident happened. Gage filled one of the holes with gunpowder and began tamping it down with the iron rod. As he was tamping it down, his tamping iron struck a spark on the rock. The gunpowder exploded, shooting the rod out of the hole like a long bullet out of the barrel of a gun. The rod hit Gage in the head, and in fact shot right through his head, entering just below the eyes and exiting at the top, to land some distance away, covered with blood and bits of Gage's brain. Most amazingly, Gage didn't die. He lived for many years after this accident, but with a large lesion in his brain. Since his case occurred in 1848, there was no brain imaging available to localize the lesion, and the reconstruction had to rely on some detective work based on his skull (which was still available), the report of the accident, and other available material. Hanna and Antonio Damasio, then at the University of Iowa, carried out a piece of detective work, finding that particular parts of the prefrontal cortex (the ventromedial prefrontal cortex, or vmPFC) were damaged in Gage's brain (Damasio et al. 1994).

The relevance to cognitive neuroscience centers on what deficits that brain lesion caused in Gage. Unfortunately, this has been somewhat unclear, since not only were there no MRI scanners back in 1848, but there was not even any postmortem anatomical analysis of Gage's brain. There was also very little available in terms of modern behavioral tests for cognitive functions. So, once again, we have to rely on reports from family, friends, and physicians that were written down at the time. There seems to be agreement that Gage's behavior was notably altered. There is much less agreement on exactly what that alteration consisted of. Gage made poor decisions in his life and clearly had some difficulties maintaining a job and keeping his former friends. Given the unique, underanalyzed and essentially anecdotal nature of this case, it will remain unclear whether these changes can be attributed solely to his lesion, or whether they might have arisen also, or perhaps only, from any of a number of other causes in his life that were uncontrolled in this case. Importantly, we do not know whether a similarly severe brain lesion anywhere else in the brain might have resulted in a similar impairment. Nonetheless, Gage's emotional behavior was altered—he has been described as becoming insensitive, profane, and exhibiting poorly controlled emotions.

This single case motivated the hypothesis that the prefrontal cortex, and perhaps especially the vmPFC (which were regions damaged in Gage's brain, although not the only regions damaged), have something to do with emotions. If we zoom forward 150 years, we then find a number of more modern lesion studies (seminal ones from the same group of the Damasios at the University of Iowa) that indeed largely support that hypothesis. These modern lesion studies had structural MRI or CT available and so could identify the anatomical primary lesion at least at the level of resolution possible with this method (which is around a millimeter or so), and they included much more extensive and standardized behavioral and cognitive tasks to assess the impairment that was produced by the lesion.

The lesions in these modern patients were typically caused by resection of a tumor, or perhaps resection of an aneurysm (the bulging of a blood vessel), which is considerably more specific than having an iron rod shoot through your head (but considerably less specific than experimental lesions in animals). While unambiguous damage to the prefrontal cortex could be verified with structural MRI, there is generally also variable damage to white matter or subcortical regions such as the basal forebrain. Perhaps in part because of the variability in location and extent of the brain damage, there is still considerable heterogeneity in the behavioral impairments produced across these studies. There have been reports of impairments in recognizing emotion from facial expressions (Hornak, Rolls, and Wade 1996), and there have been reports of dysregulated emotions like anger (Koenigs and Tranel 2007), and of blunted social emotions and empathy (Koenigs et al. 2007). Although these findings all have something to do with emotion, the evaluation of the patients generally did not focus specifically on emotion states (the above three papers, respectively, focused on judging emotion categories from facial expressions, on monetary choices made in economic games, and on moral judgments). So, just as for fMRI studies (see also chapter 8), there is tremendous heterogeneity just in what experimenters mean by the word "emotion" in their papers, making it difficult to compare across the studies and draw generalizable conclusions from them.

There are also more specific reports of altered emotional behavior and self-reported emotional experience in patients with damage to the prefrontal cortex, giving this domain of emotion more consistency than the studies alluded to above (Hornak et al. 2003; Anderson et al. 2006). Nonetheless, it remains quite challenging to assess emotions in these cases: self-report data are particularly problematic, since it is known that the prefrontal cortex is important for insight into one's own condition. In fact, despite apparently profound difficulties in real life, these patients often deny that there is anything wrong with them. Further compounding this problem, it is rare to have equivalent measures of emotion before and after the lesion, since the patients don't come to the attention of scientists until they have already had the lesion. Consequently, the power of within-subject lesion studies, which are common in animal research, is often unavailable in human studies.

It remains unclear in all these studies exactly what is going on in the brain. That is, we can see, on structural MRI scans, that there is a focal lesion in the prefrontal cortex, typically sustained some years ago. But we would like also to know what the function of the rest of the brain looks like now: have there been long-term changes elsewhere? How does the pattern of activation compare to what is typically seen in healthy people tested on the same task (or seen in the same patient prior to the lesion)? Just knowing that one region is lesioned tells us little about how this changes function elsewhere in the brain, which is what we ultimately need to know to explain how altered behavior might be generated, and what we would ultimately need to know in order to fully describe any altered central emotion states (this point applies to some extent to optogenetic studies in animals as well).

For example, it has been shown that lesions to the ventromedial prefrontal cortex cause changes in the fMRI activation within structures that are connected with this brain region, such as the amygdala (Motzkin et al. 2015). There are also the obvious limitations of sufficient control conditions, general lack of availability of within-subject data (before and after the lesion comparisons are usually not possible, since the lesions are not experimentally made, a problem to which we already alluded), imprecision and uncontrollability of the lesion itself, and capacity for compensation and reorganization in the brain following a chronic lesion. These are all big differences compared to the well-controlled studies in animals that we just reviewed. While human lesion studies remain extremely valuable, interpreting their results is fraught with challenges and generally requires corroborating data from other techniques or species.

All this might make you think that human lesion studies have limitations so severe that they preclude scientific utility, but this harsh conclusion is premature. For one thing, all methods have their strengths and limitations, and human lesion studies remain valuable in establishing the necessary role of a brain structure, a type of information that correlational methods like fMRI cannot provide. For another, it is indeed quite possible to combine lesion studies with fMRI in the very same patients, and this is likely to be a substantial source of new data going forward. It is also worth noting that many, perhaps most, of the limitations previously noted could be substantially improved in good part just by increasing sample sizes and administering more consistent tasks across patients, if an effort were made in the research community. There is no shortage of patients with strokes and tumors, just a lack of infrastructure for incorporating them into a large research consortium (Adolphs 2016).

Some studies have indeed used samples of a few hundred lesion subjects, and have used sophisticated voxel-based statistical mapping techniques to analyze the relationships between focal brain damage and specific impairments across such a group. Those studies have been able to provide a much more fine-grained, and statistically robust, set of conclusions. The take-home message is that the majority of published lesion studies should be considered with caution (especially for older studies) and rarely provide clear conclusions about the neural systems that underlie emotions—but that modern lesion studies (using large samples and newer methods, as well as combining lesions with fMRI) are substantially improving on this.

The prefrontal cortex is currently one of the most investigated regions in human emotions using fMRI. In fact, it is better suited to investigation with fMRI than is the amygdala, because the fMRI signal is typically much better in cortex than it is in the amygdala for technical reasons. The prefrontal cortex is a region of the brain that is disproportionately larger in primates than in other animals. It consists of several different sectors, each of which contribute to processing emotions (figure 8.1 only shows a very few). In humans, the functions of the prefrontal cortex related to emotions have been conceptualized as content-specific appraisals. Thus, each of the different sectors have been associated with emotion appraisals of different sorts: appraisals of exteroceptive sensations, of memories and imagined future events, of visceral interoceptive signals, and so forth (see Dixon et al. 2017, for review). This is thought to be achieved by the widespread connectivity of the prefrontal cortex; each of the different sectors implements its emotion function by serving as a hub for inputs from a network of other brain regions. For instance, ventral and anterior parts of the cingulate cortex are thought to be most closely related to connections with visceral and autonomic components of emotion, whereas lateral prefrontal cortex is thought to be most closely related to emotion regulation (see box 3.4). While still quite preliminary, such a scheme will help us eventually to understand the very complex and indirect ways in which emotion states can be induced and coordinated in humans.

In addition to the ventromedial prefrontal cortex, two other regions of the human brain thought to be related to emotion have been highlighted by lesion case studies. One of these is the insula, the second is the amygdala (figure 8.1). These two structures are of particular interest because they are really the only candidates so far for brain regions involved

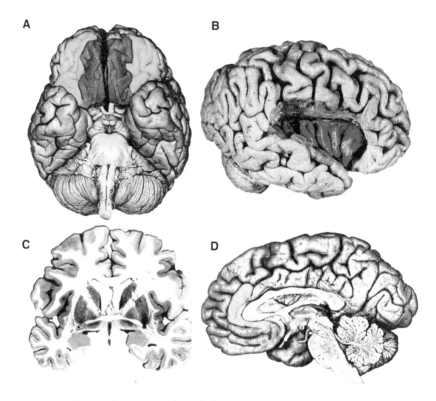

FIGURE 8.1. Brain regions commonly studied in human emotion. A: medial (red) and lateral (green) orbitofrontal cortex, which are parts of the prefrontal cortex; B: the insula, which is a cortex that represents body states, buried underneath the frontal lobe (which has been partially cut away in this image so that one can see the insula [purple] underneath); C: the amygdala, which is in the medial temporal lobe and shown here in orange on a coronal section of a brain; D: the anterior cingulate cortex (yellow), which also overlaps with the ventromedial prefrontal cortex. Lesions in each of these regions have been associated with: blunted or dysregulated emotions (A), impairment of disgust (B), impairment of fear (C), and lack of emotional arousal and motivation (D). The hypothalamus and periaqueductal gray are also involved in human emotions, just like they are in animal emotions, but are rarely investigated in part due to the difficulty of resolving them with fMRI.

disproportionately with a single, specific human emotion (Calder, Lawrence, and Young 2001). The insula accounts for about 2 percent of our cortex but cannot be seen at all from looking at a brain from the outside. It is buried deep within the brain and consists of several distinct regions. Its functions are still very much under investigation and range from processing of gustatory and visceral signals and pain to processing complex social emotions (see Nieuwenhuys 2012, for review).

The insula has been implicated in disgust on the basis of electrical stimulation, lesion, and fMRI data. It is activated across a whole range of stimuli and tasks, from tasting, smelling, or thinking about disgusting foods to watching somebody be disgusted (for instance, from their facial expression), to moral judgments involving disgust. In fact, the insula is activated across all of the different varieties of disgust that we showed in figure 1.3. This is an interesting example of neuroscience data (shared activation of a common brain region) supporting a functional hypothesis (that the varieties of disgust share functional and algorithmic properties). The link between the insula and disgust makes some anatomical sense, since the insula represents interoceptive information from the body—such as signals from the stomach that would make you feel nauseated.

Interestingly, there is a close correspondence between sensory representations related to disgust (for example, bad taste), visceral sensations (for example, nausea), and also the conscious experience of disgust—all of these activate the insula, although the more complex emotional aspects of a conscious experience activate more anterior insula regions than do the more sensory aspects. This has made the insula a structure frequently encountered in theories of the conscious experience of emotion. Some theories argue that the insula, together with medial prefrontal cortex and amygdala, is part of a network for representing, maintaining, and predicting our own body states. In particular the predictive function of this brain system has been highlighted in theories of emotional feelings (Seth 2013; Feldman Barrett 2017), emphasizing that emotions are not merely reactions to stimuli but also depend on expectations (see chapter 10).

The amygdala has been linked to processing fear, just as in the animal studies we reviewed in chapter 6, and probably constitutes the most compelling case of a brain region's necessary involvement for a specific emotion. This is not to say that the amygdala only processes fear—we know from electrophysiological and fMRI studies that many different types of emotion signals are processed by circuits in the amygdala (see chapter 6). But focal lesions do seem to produce the most severe impairments just in fear (box 8.2). Interestingly, both amygdala and insula lesions have been reported to impair both the perception and experience of emotion (fear and disgust, respectively) (Calder, Lawrence, and Young 2001). This has supported a highly influential hypothesis: that the ability to recognize emotions in other people depends on being able to feel the emotion oneself (Goldman and Sripada 2005).

BOX 8.2. Patient S.M. and dissociations of fear.

Perhaps the most famous modern patient in the neuroscience of emotion is a woman called S.M., whose emotional life has been studied in great detail (Feinstein et al. 2011; Feinstein et al. 2016). S.M. came to the attention of Daniel Tranel and Antonio Damasio at the University of Iowa around 1990 because she had an extremely rare brain lesion: a bilateral lesion of the amygdala. The lesion was caused by a genetic disease, Urbach-Wiethe disease, which results from a mutation in the gene coding for a protein called extracellular matrix protein-1 (but nobody has any idea how this gene mutation causes the brain lesion).

S.M. had an unremarkable overall neuropsychological profile. Her intelligence, memory, vision, language, and other basic perceptual and cognitive abilities were all broadly in the normal range. She did, however, show highly selective impairments in domains related to processing fear. For example, she did not show Pavlovian fear conditioning of skin-conductance responses (even though she had declarative memory for US-CS associations); she had difficulty recognizing facial expressions of fear (but could recognize facial expressions of other emotions); and she did not appear to feel fear (but experienced other emotions normally).

There have been two particularly important results from studies with S.M. One result has been a hypothesis about a brain mechanism, mediated in part by the amygdala, by which we recognize emotions from facial expressions. This hypothesis proposes that allocating visual attention to specific facial features (notably the eye region of the face) is one component enabling our ability to judge fear from facial expressions. S.M. was impaired in recognizing fear in facial expressions, but also did not spontaneously make use of information from the eye region of faces, and indeed did not look at the eyes in faces (figure 8.2). Instructing her to look at the eyes in faces indeed improved her ability to recognize fear, support for a causal mechanism whereby amygdala lesions impair fear recognition because they impair a mechanism for attending to fear-relevant features in faces (the wide eyes) (Adolphs et al. 2005a).

A second important result from the study of S.M. has been the demonstration of specific dissociations that help us to make scientifically principled distinctions. As already noted, S.M. is disproportionately impaired in the domain of fear, relative to all other emotions, which is compelling evidence that fear is a distinct emotion category. Another very important dissociation shows that there are some varieties of fear that appear to be intact, or even exaggerated, in her. For example, although she is impaired both in her ability recognize fear in faces and in experiencing fear from horror movies or other external stimuli, she experiences panic induced by inhaling carbon dioxide (Feinstein et al. 2013). This causes a sensation as if one were suffocating, and about a quarter of healthy people experience a panic attack. S.M. and two other patients with bilateral amygdala lesions both experienced severe panic when they inhaled carbon dioxide, demonstrating that the panic state resulting from suffocation signals does not require the amygdala. Thus, exteroceptive fear and interoceptive panic appear to be two neurobiologically distinct states of "fear."

Another key dissociation shown by S.M. is that she has a mostly intact concept of fear, even though she cannot recognize fear in faces, and even though she does not experience fear. When you ask S.M. what fear is, she can tell you that you would feel afraid if chased by a bear, or that people who are afraid tend to scream and run away. She can use the word *fear* appropriately in conversation. One reason for her apparently intact concept of fear may be that, in humans, we rely primarily on language to assess somebody's concept of fear. S.M. has intact language abilities and intact declarative memory for consolidating semantic memories of facts that can be represented lexically. She reads books, she watches movies, and she talks to other people. So she will have acquired much of the conceptual knowledge about fear that people can talk about

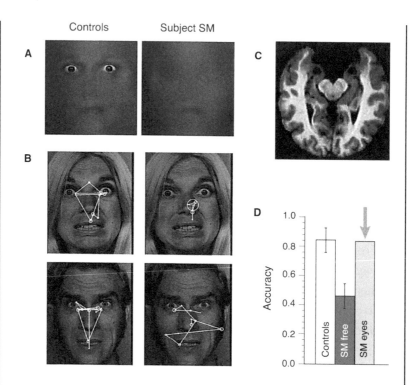

FIGURE 8.2. Amygdala lesions impair the processing of information from the eye region of faces. Data are from patient S.M., who has complete bilateral amygdala lesions and is impaired in recognizing fear. (a) Regions of the face that subjects use to recognize emotions. Patient S.M. differed from controls in making much less use of the eye region of the faces. (b) While looking at faces, S.M. (*right*) made fewer fixations to the eyes than did control subjects (*left*). White lines show the eye movements made by subjects as they looked at these faces. (c) S.M.'s amygdala lesions can be seen in an MRI scan (the round black circular regions near the top of the image). (d) When S.M. was instructed to look at the eyes ("SM eyes") in a whole face, this resulted in a remarkable recovery in her ability to recognize the facial expression of fear ("SM eyes," arrow) compared to her accuracy prior to this instruction ("SM free"). Modified with permission from Adolphs et al. 2005a.

and think about in language, even without experiencing fear herself. These dissociations are important, because they show us that conceptual knowledge about emotions can be largely separated from actually having the emotions, or recognizing the emotions from nonlexical stimuli like facial expressions. The concept of an emotion, the conscious experience of an emotion, and the emotion state itself are three different things.

Another approach to establish causal relations between a brain region and emotion is electrical stimulation. This can be done in human neurosurgical patients under some circumstances and has many of the same limitations as the lesion studies: the sample sizes are very small, and the extent and location of the stimulation is poorly controlled. Consequently, the effects are typically quite variable. Nonetheless, as with the lesion studies, occasionally one does encounter remarkably specific effects: electrical stimulation of the basal ganglia can cause intense sadness, and stimulation of the premotor cortex can cause mirth and laughter, for example. And classic electrical stimulation studies of the insula have produced disgust or nausea-like sensations, whereas electrical stimulation of the amygdala has produced thoughts, memories, and subjective experiences related to fear. Yet, as we already noted in the previous chapters, these findings are difficult to attribute to selective activation of neurons in the insula, or in the amygdala, since large regions of brain tissue are stimulated, including both neurons and axons.

Nonetheless, when surveying the literature from electrical stimulation studies, three broad conclusions seem apparent (Guillory and Bujarski 2014). First, aspects of emotion (typically reports of conscious experiences of an emotion) can be evoked from a number of different cortical and subcortical regions. Second, there does appear to be some specificity, and many of the same regions we already encountered (insula, amygdala, periaqueductal gray) can evoke emotions when stimulated. However, this conclusion is qualified by the fact that there is a severe sampling bias when neurosurgeons implant electrodes and stimulate—the brain regions to stimulate are not chosen randomly or homogeneously. Third, there also is clear specificity for certain emotion categories. Some stimulations evoke happiness, some sadness, some disgust, some fear. While the precise categories of emotions that are evoked are somewhat unclear and the evidence is typically sparse, it is not generally the case that stimulation of one and the same location evokes all kinds of different emotions.

Similar studies with larger subject samples and with better control conditions are possible using techniques such as transcranial magnetic stimulation (TMS). However, TMS has rather poor spatial resolution. There is ongoing work on developing safe perturbation methods that can be applied to human brains with better spatial and temporal specificity (for instance, disruption using focused ultrasound). The tools available for perturbing human brain function may thus look a lot more precise in the near future—however, they still are unlikely to distinguish specific neuronal populations any time soon, since optogenetic or pharmacogenetic studies are so far limited to animals for ethical reasons (box 5.5).

Given the poor anatomical specificity of both the lesion and electrical stimulation studies, poor control over confounding variables, and given the typically small number of subjects, it may seem difficult to know what to conclude from them. We would draw two main conclusions. First, although there is enormous variability, even single cases can be sufficient to demonstrate specific causal relationships, or to logically show us dissociations that are possible. One conclusion from lesion studies is that specific, individual emotions can be compromised relatively in isolation, at least for the case of fear, and to some extent for the case of disgust. Pinning this specifically to the amygdala and insula is more problematic, given inadequate anatomical resolution, but it is also not necessary in order to conclude that there is *some* anatomical segregation for processing individual emotions in the brain. Often, people have derived a number of much stronger conclusions, such as that the amygdala is *only involved in processing fear*, or that *only the amygdala* is involved in processing fear, but these are entirely unwarranted and depend on a flawed logic, as we already pointed out in chapter 4. So, although the conclusions we can draw are both anatomically and conceptually limited, the data do show that there are specific brain systems that may be necessary or sufficient for specific components of certain emotions in humans.

It is noteworthy that this small number of classic lesion and electrical stimulation studies from humans has largely fueled the specific anatomical hypotheses about emotion that are still tested today, using fMRI in humans, and using optogenetics, calcium imaging, and other methods in animals. Thus, although the amygdala represents only one of hundreds or thousands of anatomically distinct brain regions, the vast majority of studies of the brain in emotion (in both humans and animals) have focused on the amygdala. An important future direction would be to investigate the neural systems for emotion with a less anatomically biased lens—something that whole-brain fMRI is now routinely accomplishing.

This is an important point: if you search the literature, you will find strong apparent evidence that specific brain regions are involved in emotion, such as the prefrontal cortex and the amygdala. But this may simply reflect a strong bias that these are the main regions where experimenters have actually looked in the brain, or what they have chosen to publish about. Such biases are well known and need to be corrected if one does simplistic meta-analyses of published studies.

For instance, one modern tool for mining the fMRI literature, a program called "Neurosynth" that was developed by the cognitive neuroscientist Tal Yarkoni, lets you visualize where in the brain there is activation across published papers that have certain key words in them (Yarkoni et al. 2011). If you type in "emotion," you get 790 studies in the database (no doubt a considerably larger number by the time you read this book) reflecting over 28,000 reported fMRI activations that surpassed some statistical threshold worth reporting. All these papers had the word "emotion" in their abstract and had something to report about emotion. Where are these activations concentrated, if we pool all the 790 studies together? Figure 8.3 gives us the answer: mostly in the amygdala! Needless to say, this is basically just a summary of the literature's bias in investigating the amygdala.

Although figure 8.3 indicates some brain regions that seem to be reported as activated (that is, correlated) relatively indiscriminately for "emotion," it is noteworthy that no structure has yet been found that is necessary for processing all of emotion. All the impairments found with lesions thus far seem to impair components of specific emotions, or to dysregulate components of emotions, but not to abolish all emotions altogether. Thus, patients with insula lesions, or with amygdala lesions, still appear to have emotions—even though specific ones (like disgust or fear, respectively) may be impaired in certain ways. This is quite an important finding that has been used to draw two main conclusions. One conclusion that has been drawn is that there are multiple emotion systems in the brain (so you would have to lesion all of them to completely abolish emotion, and the neuroanatomy is just too distributed for that).

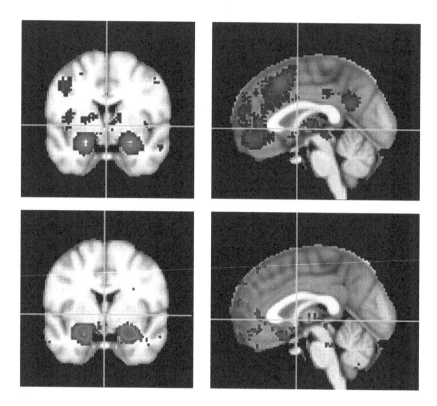

FIGURE 8.3. Meta-analytic mapping of brain activations for emotion. Using the program Neurosynth, a very simple search for the keyword "emotion" yielded results from 790 fMRI studies. These studies all contained the keyword "emotion" and were about emotion in some way—but they were extremely varied. In the top panels (in blue) are the activations found when you ask where all activations (above a statistical threshold) in these studies having to do with "emotion" are located, the so-called forward inference map. This prominently shows the amygdala (*top left*) and prefrontal cortex (*top right*). However, many of these regions may also be activated in many other studies that do not have anything to do with emotion. A more specific analysis thus asks which regions were activated by those studies containing the keyword "emotion," but not in any other studies in the database (that did not contain the word "emotion"). This so-called reverse-inference map is more useful and shown in the bottom panels (in red). The images are screen captures generated online in a few seconds through www.neurosynth.org and show the activations superimposed on a standard brain template. Researchers use Neurosynth for more complex kinds of inferences and can map brain activations not simply to single words but to topics of related words as well.

A second very interesting conclusion that has been proposed is that it might be possible to completely abolish all emotion in principle—but only by putting the person in a coma. That is, it may be that there are specific brain regions that would abolish all of emotion, but that these cause death or unconsciousness, so that it just turns out to be impossible to have a conscious person who has no emotions at all. This of course could give us some interesting insights into the functional role of emotion, and some links to conscious experience, which are indeed insights that have been hypothesized by several people and which we discuss in chapter 10. For the rest of this chapter, we will now leave lesion or electrical stimulation studies and focus instead on neuroimaging.

fMRI Studies of Emotions: The Method

By far, the majority of neuroscience studies on emotion in humans use neuroimaging, and they now number in the thousands. Some earlier studies had used positron emission tomography (PET), but this is rarely used nowadays because of major limitations in resolution, and because it requires administering radioactive materials to the subjects. Nowadays, neuroscience meetings on human emotion primarily feature fMRI (functional magnetic resonance imaging) and electroencephalography (EEG), with a few using other techniques such as magneto-encephalography (MEG) or transcranial magnetic stimulation (TMS). Each of these methods has strengths and weaknesses; we gave a brief summary in figure 4.2.

A primary strength of fMRI is that it is relatively easy on the subjects—unlike PET, you do not need to administer radioactive materials, for instance, and it is entirely noninvasive with no known long-term side effects (provided you don't have any metal in your body, since the method involves being in a very strong magnetic field). In a typical experiment, subjects lie on their backs in the scanner for about an hour while brain data are collected. The only drawback is that you need to keep still for that period of time, and you may well get sore or bored by the end of the hour. Unfortunately, fMRI is not particularly easy on the researchers—a modern MRI scanner will cost between $2 and $10 million to buy, will cost several hundred thousand dollars a year just to operate, and will require specialized technicians. It is so complicated a technique that nobody can build one in their own lab or figure out how to analyze the data entirely by themselves. The only way to have fMRI as a method is to have a big market, and a large user community that can share knowledge—which are both indeed the case.

fMRI has reasonably good spatial resolution, close to a millimeter, and reasonably good temporal resolution, about a second. "Reasonably good" here is of course quite relative: human lesions, by comparison, have spatial resolutions typically on the order of centimeters and no temporal resolution at all, but single-unit recordings have single-neuron resolution with millisecond precision (see figure 4.2). A major limitation of fMRI is that it has very low signal-to-noise, and consequently a lot of data and a lot of fancy processing are required to extract the signal from the noise. One reason for this is that the fMRI signal does not directly reflect electrophysiological activity. Instead, it is a measure of the local change in the ratio of oxyhemoglobin to deoxyhemoglobin in the blood supply to the brain, resulting from changes in blood flow to a region. This change in the local oxygenation of blood causes very small changes in the magnetic field that the fMRI method can detect. So, one immediate concern is whether this indirect hemodynamic measure actually reflects electrophysiological activity at all. It turns out that, most of the time, it does (but it need not always).

The next question is, what aspect of electrophysiological activity does it reflect? Changes in blood flow in the brain happen so that those regions that are "working hard" (because they are carrying out some cognitive process that we are experimentally manipulating) can have an increased supply of glucose and oxygen. "Working hard" means "have high metabolic demand," which corresponds mostly to biochemical events at the synapse, the structure where neurons communicate with one another. Consequently, the fMRI signal is best correlated with field potentials and synaptic processing (including inputs to a region), but not with action potentials in projection neurons that transmit information out of a region (which is less metabolically demanding). This is an important point to keep in mind, since it means that single-unit recordings could be relatively unrelated to the fMRI signal from the same region. Indeed, a strong fMRI *activation* could quite well reflect a strong *inhibition* within a region through local inhibitory interneurons; conversely, a strong fMRI de-activation could quite well reflect a release from such cellular-level inhibition.

Note that "activation" and "de-activation" in fMRI are always relative to some contrast or baseline. There is no such thing as a pure activation, since any baseline level in the brain never corresponds to zero activation—the brain is active all the time and has high metabolic demands and blood flow all the time. If there were no metabolic activity or blood flow in the brain, you would be dead. Ironically, you can actually find some fMRI signal in a dead brain, if you decide to do a somewhat silly analysis using flawed statistical methods. One study reported an fMRI activation in the brain of a dead salmon, winning an IgNobel prize for the researchers' efforts, and highlighting the careful statistical corrections that are required to avoid false positive results in fMRI studies (corrections that were not done in the study, to show how you can get spurious findings).

Related to this funny finding of an fMRI response in the brain of a dead salmon, the low signal-to-noise and multivariate nature of fMRI has in fact resulted in some crises that the field of neuroimaging has weathered: past studies were often underpowered or used inappropriate statistical approaches, resulting in published studies that did not replicate. For example, great care needs to be taken to correct for multiple comparisons in statistical tests, and the multistep nature of fMRI data analysis needs to ensure that no statistical dependencies creep in such that one artificially selects specific subsets of the data that show the highest statistical significance. However, fMRI is such a huge field, with such a large number of investigators, that these problems, once identified, resulted in considerable efforts to correct them. In fact, as a whole, the field of human neuroimaging is remarkable for how it has managed to solve many very difficult problems of finding small signal in high-dimensional and noisy data. One current direction is to build consortia, or data-sharing platforms, where very large sets of data can be accumulated, so that these can be mined with sophisticated statistical methods and adequate power.

A popular approach to the limitation of poor signal-to-noise, an alternative to conducting a new analysis on a larger sample size, is to use already collected data from prior studies and conduct meta-analyses. Such meta-analyses have become fairly popular in affective neuroscience. However, the results that these analyses produce are always dependent on the quality and homogeneity of the studies that go into them in the first place. In figure 8.3 we didn't pay attention to any of that—and got a misleading picture that you might interpret as showing that "emotion is in the amygdala." Although there are now several sophisticated meta-analytic techniques available, and although the number of fMRI studies on human emotion available for a meta-analysis is in the thousands, there are as yet few or no convincing meta-analytic results for the simple reason that the individual studies are just too heterogeneous and often of too poor quality.

For instance, one large meta-analysis of emotion failed to find any reliable association between activations in any brain regions, and specific emotions (Lindquist et al. 2012)—likely in part for the reasons just mentioned. This is particularly likely given that specific emotions are typically studied with specific induction methods. Thus, one study might investigate "fear" using horror movies, and another might investigate "fear" using pictures of spiders, and yet another might investigate "fear" using a variety of autobiographical memories. In fact, the situation is much worse than that, since meta-analyses often pool together studies in which people are asked to recognize fear in facial expressions, think about a fearful situation, or made to experience fear themselves. As you can imagine, the brain states produced by these different types of tasks are hugely different. They all involve some processing related to emotion, but only in the broadest and most derivative sense. It is quite implausible that the same type of functional state, let alone brain state, is involved across all the different types of study, even though the researchers might use the label "fear" somewhere in their paper.

However, future meta-analyses on emotion will certainly be capable of producing more interpretable results. Both the number, and the quality, of fMRI studies on emotion keeps increasing, and researchers might heed the distinctions that we emphasize in this book and not lump together everything labeled "emotion." There is also an increase in standardization of methods and analysis pipelines. With more sophisticated filters (not lumping together heterogeneous studies, and carefully filtering out studies of poor quality), and more sophisticated statistical tools, such future fMRI meta-analyses will be a good test bed to see if fMRI is fundamentally capable of finding emotion systems in the human brain—or if these are just at a level of resolution below what is possible with this technique. Such meta-analyses will also be helped by the further development of tools for easy data sharing, such as the OpenfMRI project being developed by Russ Poldrack at Stanford University.

There is one feature of fMRI that is a very major plus, and a respect in which it definitely surpasses single-unit recordings or calcium imaging. That is its whole-brain field-of-view. This makes it possible to go well beyond searching with neuroanatomically narrow hypotheses for activation just in the amygdala, or in the insula. One can look at the whole brain at once, including all of the cortex and all subcortical and brainstem structures. A modern fMRI study can now acquire activation data from an entire brain about once a second. That's a lot of data! You can thus think of the fMRI data as a time series, sampled at about 1 Hz, that measures the fMRI signal at each of a million little cubes (called "voxels"; each 1 mm voxel has about 10,000 to 100,000 neurons) in the entire brain. However, that large field-of-view creates some analysis challenges of its own; we need to correct for many multiple comparisons. At a standard statistical alpha = 0.05 threshold, we would find one "significant" voxel in every 20 or so voxels just by chance. That would amount to about 50,000 voxels all over the brain, and no doubt some of them would cluster in some interesting brain regions—but they could be false positives. In this respect, the multivariate nature of fMRI is similar to the challenge posed by genome-wide association studies, where one also has to control for the many different genes that are probed for a significant effect.

As we already noted, these challenges have been very actively addressed by the neuroimaging community, and the whole-brain field-of-view of fMRI, treated carefully, is certainly not a liability (after all, you can always choose not to use it and instead focus your analysis on a single brain region, a so-called region-of-interest, or ROI, an approach analogous to looking at a single candidate gene). The whole-brain field-of-view allows us to see, all at once, activations anywhere in the brain—and, to some extent, to infer how they might be related to one another by looking at their correlations. This also opens up several additional types of derived measures. There is now a strong focus on network-level analyses of fMRI data that focus less on activation in particular regions than on the functional connectivity between regions inferred from temporal correlations. Indeed, functionally coupled networks can be derived in this way even from so-called resting-state fMRI, in which subjects don't do any task at all but just lie passively in the scanner while the spontaneous (but partly correlated) fluctuations in the fMRI signal are measured everywhere in the brain. This yields a large correlation matrix, from which in turn a large number of network metrics can be derived. Similar to studies of social networks or of the internet, one finds that not every region is connected with every other region. Instead, there are "hubs" that coordinate a lot of the connections between regions, much like internet servers and routers connect different computers.

Given the highly multivariate nature of the fMRI data, one can use multivariate analyses of activation across many voxels; that is, one can look for *patterns* of activation, rather than mean signal changes, in the brain. Given the whole-brain field-of-view, one can search all over the brain without requiring an anatomical hypothesis a priori (this requires careful statistical correction). In a nutshell: despite its fundamental limitations in being an indirect hemodynamic metric, having modest spatiotemporal resolution, and having low signal-to-noise, fMRI is in fact a very rich measure that, with enough data and with sufficiently sophisticated tools, can be used to probe brain function not only in terms of overall average activation within a brain structure, but also in terms of the detailed pattern of activation within that structure, or in terms of the functional connectivity of that structure with other regions in the brain.

This is a very different kind of measure of brain function than recording from single neurons, as is often done in animal studies and as can be done in humans also in a clinical setting. It raises interesting questions about the best method to use. In chapter 6, we saw that specific neuronal subpopulations may be the most functionally meaningful unit of analysis (rather than single neurons, or specific anatomical brain structures). fMRI may reveal to us more counterintuitive brain measures related to multivariate, connectivity-based metrics, but it is possible that these will be more useful functional units by which to understand how emotion is processed in the brain. Rather than searching for specific brain regions or cell populations involved in processing emotions, we may end up finding complex, distributed patterns and networks that are the fundamental units by which to understand emotion processing.

There are new approaches made possible just by the sheer amount of data available. If you have a lot of data, you can choose to analyze only a part of your data, and then test the results of your analysis to see if they can predict the remainder of the data—that is, you can take an approach that shows whether results replicate and generalize. Such a predictive framework is now being used in many fMRI studies. Rather than simply analyzing all of your data and deriving, for example, a correlation coefficient whose magnitude describes the observed effect size in your particular dataset, it is possible to use only a portion of the data to fit a model that predicts a data point not used in the construction of this initial model. That data point can then be used to test how good the model is. Modern approaches using techniques from machine learning are now widely used in fMRI analyses, and they typically feature such a predictive framework, which is considerably more powerful and reliable (since it makes out-of-sample predictions) than traditional approaches (Dubois and Adolphs 2016).

Using such predictive approaches, and multivariate measures, has radically changed fMRI data analysis in the past few years. It now offers much greater sensitivity to detect effects, and much greater reliability and replicability once they have been detected. One key ingredient that is widely recognized to make all this work is larger sample sizes, and sharing of data. For instance, one of the most widely analyzed datasets comes from the Human Connectome Project (http://www.human connectomeproject.org/), which features over one thousand subjects, each with several hours of the highest quality fMRI data. Such sample sizes are not currently feasible with lesion studies or single-unit recordings, making fMRI by far the best approach for producing results that will generalize. (On the other hand, compare the resolution of the human brain in the Human Connectome Project with that possible using microscopic analysis of brain structure in postmortem brains, as is being done at the Allen Institute for Brain Science: https://twitter .com/Allen_Institute/status/776492924942942208.)

Although powerful, training machine learning algorithms to classify patterns of brain activation does not provide any understanding of causal mechanism by itself. The approach may well have diagnostic and prognostic value for mood disorders, for example, but it would fall short of allowing us to design new interventions to treat such disorders. Comparisons with studies in animals will always be essential in this regard (see chapter 5). Nonetheless, it turns out that it is actually possible to derive some insight into causal mechanisms using fMRI, a very new development (box 8.3).

BOX 8.3. Causality from fMRI?

Given everything we discussed in chapters 4 and 5, you might think that it is absolutely impossible to derive any causal conclusions from fMRI data, which are inherently correlational. However, this is not quite the case, for two reasons. One obvious reason is that it is possible to combine fMRI with causal interventions. For instance, one can activate a specific brain region through electrical stimulation or optogenetics while measuring fMRI elsewhere in the brain to visualize the causal effects of that stimulation. This approach has been taken in rodents, monkeys, and humans (see figure 11.2). Indeed, it has been used to begin mapping out the functional connectivity of the human amygdala in humans (Dubois et al. 2018). An important caveat with these kinds of studies is that, while they do show causality, it is much trickier to conclude anything about how direct these causal effects are. Activation of the amygdala produces robust fMRI activations all over the brain, but the majority of these are going to be very complex network-level effects, not direct monosynaptic inputs from the amygdala.

However, it is possible to model causal effects even from standard fMRI data, even from resting-state fMRI data, without any intervention at all—provided that the assumptions of your model are met (which is almost always an idealization). One way of doing so uses temporal relations between the multiple activations and experimental stimulus presentations that are occurring during an fMRI experiment; since causes must precede their effects in time, the ability to predict activation in one brain region from preceding

---

activation in another could in principle be used to infer causal relations. A number of techniques capitalize on this idea, including Granger causality and dynamic causal modeling. One problem with all of them is that the temporal resolution of the actual fMRI signal is very poor, and much lower than cause-and-effect events among neurons (seconds versus milliseconds), requiring complex models and assumptions to make this approach work.

Another approach ignores temporal order altogether and simply looks at which fMRI activations happen to be correlated. As we saw in chapter 4, finding a correlation between two brain regions, A and B (figure 4.1a) means either that A caused B, B caused A, or both A and B were caused by something else (figures 4.1b, c, d, respectively). So there are three possible causal structures that are all consistent with observing the correlation. Can we eliminate some of them with further data? If we now also measure activation in region C, and we find that it is correlated with both A and B, then this would be further evidence that A and B might be caused by C (figure 4.1d). If we now regress out the activity in C (that is, calculate the correlation between A and B when there is not also a correlation with C), we might find that the partial correlation between A and B becomes zero. If that happened, we could indeed conclude that A and B are correlated only because they are both caused by C, but that there is no direct causal effect of A on B or of B on A.

This kind of logic, extended to the many brain regions from which fMRI gives us measurements, can indeed give us strong evidence of causal structure just from correlational data—provided that you measure all of the possible causes. Luckily, fMRI does a pretty good job of giving us such a complete set of measurements, since it measures activations from the whole brain (but, of course, there are inputs from the senses and from the body that we still need to take into consideration). Another important assumption is that the causal arrows are all directed one way. If we have feedback loops, things get more complicated—and we know that there is indeed feedback everywhere in the brain. More sophisticated causal discovery algorithms are being developed to handle very large datasets and to incorporate issues of feedback, and it is likely that this approach will be a powerful complement to standard analyses of fMRI data in the future (see Dubois et al. 2018; Eberhardt 2017). While many different approaches are being developed for inferring causality from fMRI data, it remains the case that they all make assumptions that are not realistic—for the time being, none of them provide the strong sense of causality from experimental intervention that we reviewed in chapter 4.

Similarity Analyses

Multivariate analyses have also provided a boon to investigating similarity spaces. In chapter 3, we noted the similarity structure of emotions; emotions are often depicted in a two-dimensional space of valence and arousal, and can always be represented as more or less similar to one another. This similarity should be reflected in the brain in some way. For instance, one would expect that two instances of fear that are elicited by very similar stimuli, and that cause very similar behaviors and subjective feelings, should correspond to rather similar patterns of neural activity, compared to, say, an instance of fear and an instance of happiness that each have very different inducers, behaviors, and feelings. (This basic logic depends on an assumption philosophers call "supervenience": that any change at the mental/functional/psychological level requires a change at the neurobiological level. You might have two different brain states that are identical mentally, but you can't have one and the same brain state with two different mental states.)

Despite the limited resolution of fMRI, similarity analyses actually work remarkably well and have become a popular tool for investigating emotions. To conduct such an analysis, researchers often use "representational similarity analysis," or RSA, which has become relatively commonplace in fMRI studies, and also in some single-unit studies. While initially the application of RSA was focused on object recognition, it has now been applied to many other kinds of representations, including representations of emotions. Figure 8.4 shows the basic logic of RSA. The overarching goal is to construct at least two (and often more) similarity spaces that can be compared. One space maps the pattern of evoked neural activation, and another space maps the stimuli, or behavioral judgments. In the case of object recognition, we could thus compare the pattern of brain activation evoked, say, when you see a hand or an umbrella, with the psychologically judged similarity between these objects (or their physical visual similarity).

In terms of the neuroimaging data, the more multivariate the data (the greater the number of voxels in your multivariate measure) the more fine-grained the distinctions between the activation patterns that you can make. RSA has been very popular, and very powerful, in understanding object recognition in the human brain. Indeed, similarity relationships revealed with RSA applied to fMRI data show a good match to the similarity relationships between people's judgments of objects. More than that, each individual subject's somewhat idiosyncratic similarity judgments of objects are mirrored by the RSA of their fMRI data. Even more than that, it is possible also to predict what the neural activation pattern should look like for a stimulus never before seen; that is, you can decode the stimulus from the pattern of brain activity.

The discussion of similarity analyses so far has disregarded neuro-anatomical details and treated the "neural activation pattern" as anatomically nonspecific, or as global over the whole brain. But in fact, one can be anatomically precise. One can pick a specific brain region (say, the amygdala), and ask whether the multivariate pattern of activation within that region encodes a particular dimension of a similarity space. Or, in the absence of an anatomical hypothesis, one can take a small sphere and move it like a searchlight throughout the entire brain to search for such anatomical regions. For example, one recent study found that the multivoxel pattern of activation in the amygdala represented the dimension of valence (pleasantness/unpleasantness), at least when odors are used as the stimuli (Jin, Gottfried, and Mohanty 2015). By contrast, the average (univariate) level of fMRI activation in the amygdala has been associated with the intensity of odors, rather than with their valence (Anderson et al. 2003). This gives us one preliminary example of how different types of emotional information might be multiplexed even within the same brain region: overall activity might

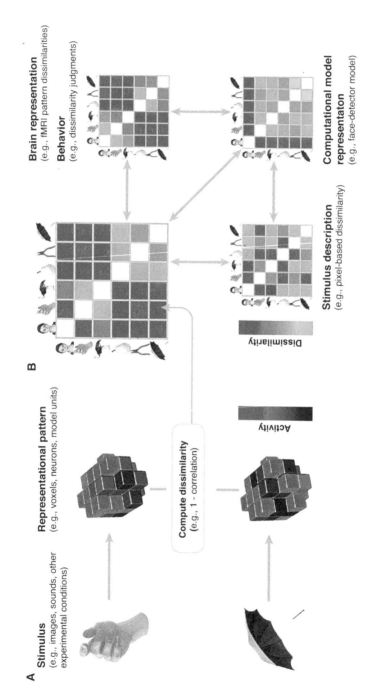

FIGURE 8.4. Representational similarity analysis. In this schematic, stimuli (A, *left*) such as seeing a hand or an umbrella cause patterns of brain activations. One can measure such activation patterns in the brain with fMRI ("representational pattern") and now calculate their similarity (or, in the schematic, dissimilarity). B. These dissimilarity matrices quantify how similar (in the brain) different stimulus representations are from one another. Similarity in brain activation patterns can in turn be compared to the actual similarity of the visual stimuli in this example, or to the psychological similarity that people give in ratings (such as ratings for facial expressions of emotions, see figure 9.7). This approach works remarkably well and often yields insightful findings. Reproduced with permission from Kriegeskorte and Kievit 2013.

encode one dimension, whereas patterns across cells might encode another. It also holds out some hope from the results we reviewed in chapter 6, where we noted that amygdala neurons encoding positive and negative valence were intermingled. That's a problem if your fMRI study just looks at the overall signal, but it may be possible to resolve using RSA (if a number of lucky assumptions are met).

This chapter spent a little time reviewing current fMRI methods (and there are many more!) to make two important points. First, the vast majority of historical fMRI studies of emotion are of problematic quality, making it difficult to derive any reliable conclusions from them. They tend to be vague about which aspect of an emotion the study is being investigated, and they tend to be statistically underpowered or problematic in the analysis. Second, the future of fMRI studies on emotion actually looks very bright, because neuroimaging has evolved so rapidly. Current fMRI studies on emotion are using bigger sample sizes, much better scanners with more powerful pulse sequences that give considerably better signal-to-noise, and correct analysis methods. Furthermore, current studies are beginning to use sophisticated multivariate measures that can yield results no standard (univariate) contrast could provide. Given that, no doubt, most studies on human emotion will continue to come from fMRI, this is basically good news for the future. Nonetheless, a final word needs to stress the importance of using additional methods to probe human brain function. We began this chapter with an overview of lesion and electrical stimulation studies, and these, as well as modern perturbation techniques still under development, will be an essential complement to fMRI for human studies of emotion.

fMRI in Animals?

We close this chapter with an important question already raised at its beginning: what neuroscience approaches can provide the best bridge between human and animal studies of emotions? Chapters 6 and 7 provided a wealth of detail about emotions in rodents and invertebrates, none of which used fMRI. This chapter (and the next) focus on humans where the modern method is almost exclusively fMRI. Since optogenetics cannot be done in humans (box 5.5), and since methods such as single-cell electrophysiology are rare and laborious, it would be a major boon to the field if we could do fMRI studies in animals.

While fMRI in rodents has severe limitations of resolution and signal-to-noise (but can be done; see figure 11.2), there are a number of fMRI studies in monkeys and other mammals, potentially providing a powerful comparison with human brains. The monkey visual system has been studied in great detail, and shows many similarities to the visual system of humans. Like humans, monkeys see in color, recognize and categorize objects, detect social signals such as direct eye gaze, and have brain systems specialized for processing faces. Our Caltech colleague, Doris Tsao, has been a leader in this field, dissecting the brain regions and the computational processes that permit monkeys to recognize each other from their faces. Based on this large literature, one would conclude that monkeys see the world pretty much like humans do. But do they understand what they see? And does what they see evoke similar emotions? Here we know that there will be differences. A monkey can certainly see a newscast on TV, but will understand none of the deeper meaning.

One interesting fMRI study compared what happens in the brains of monkeys and of people when they watch the same video in the scanner (Mantini et al. 2012). As expected, both species showed very similar activation in early visual cortices: the monkeys saw what the humans saw. However, higher brain regions showed divergent activation, presumably reflecting the fact that the humans were able to make inferences from the video of which the monkeys were incapable. Visualizing these activation differences in the brain gives us a fascinating window into how the brains and minds of these two species may be similar, but only up to a point.

A similar approach has actually been taken also to study the differences between two different people. Uri Hasson at Princeton University has examined the correlations in brain activations when people watch a movie, but he has looked at correlations not between different regions in the same brain, but rather between the same region across different people's brains. To the extent that we find such correlations, we can infer that different people process the movie similarly; to the extent that there are differences, we can infer that they do not. The emotional associations and experiences evoked by watching a movie would show a bit of both. Amazingly, this approach can be extended even to investigating how emotional information is socially transmitted; similar brain activations are obtained not only when two different people watch the same movie, but even when one person causes a brain activation pattern in another person merely by telling them what they remember about having seen a movie! (Zadbood et al. 2017). Extending such studies to better quantify the similarities and differences in brain activations to emotions will be important to fully map out variability across different species, and across individuals within a species.

Coming back to emotions in monkeys, there is good reason to believe that monkeys have at least some of the same emotions as do humans, and moreover that these depend on similar neural systems. The amygdala has been implicated in defensive emotions across rodents, monkeys, and humans, for example (Adolphs and Amaral 2016). Yet the specific stimuli that cause amygdala neurons to fire differ, and the specific behaviors caused by amygdala activation also differ. While coarsely similar types of emotion states may be present across the species, the links to stimuli and behaviors need to be interpreted relative to a given species' niche.

Aside from monkeys, what other animals could participate in fMRI studies? It would be fantastic if we could image great apes like chimpanzees, but unfortunately this is currently impossible because the scanner environment is just too aversive, and ethical considerations preclude the possibility of training the apes. It is possible that future generations of scanners could be built to accommodate apes, and a handful of neuroimaging studies have indeed been done in chimpanzees—but these have used PET rather than MRI, and the animals were anesthetized to make it work.

There is a species with rich emotional behaviors that can be trained to lie still in the scanner, however: dogs. Dogs are a particularly fascinating species for the study of emotion, not only because they evolved from a highly social species to begin with (they evolved from an ancestor of wolves about 15,000 years ago), but also because they provide a great example of two emotion properties (figure 3.2): generalization and social communication.

Studies of dog cognition have emphasized both their capabilities and limitations (Engle and Zentall 2016). Dogs (unlike wolves) excel at attending to social cues from their owners, form strong and lasting bonds with people, and generally show reduced aggressive behaviors—a genetically enabled package of behaviors, since wolf pups raised among people just like dogs don't develop these behaviors. On the other hand, many cognitive abilities of dogs presumed by their owners have not been found in careful experiments. Anecdotes claim that dogs will try to get help if their owner suddenly collapses with a heart attack, but studies find no support for this behavior without explicit training. Dog owners also tend to believe that they can tell if their pet feels guilty—but they are actually at chance in recognizing whether the dog did something wrong or not. However, the experimental studies often involve artificial situations that do not quite correspond to the real world, making it premature to draw strong conclusions from them.

Work by Greg Berns at Emory University has begun to investigate the dog's brain using fMRI (Berns 2017). Once dogs are trained to remain still, they can be given hearing protection and lie in an MRI scanner while performing tasks just like human subjects (figure 8.5). Many of the same sets of brain regions that are activated reliably in simple human fMRI studies are also found in dogs. Regions for smell and hearing, and even visual regions that seem to be activated disproportionately by faces have been reported. While still preliminary, more complex cognitive functions have been investigated in the scanner as well. For instance, fMRI responses to people and rewards have been used to assess suitability for being a service dog, and activations during a go-nogo task has been used to study the ability of dogs for self-control.

There is little doubt from their behavior that dogs have emotions—they show strong evidence of states that exhibit the properties we listed in figure 3.2. What is far less clear is what to call these emotions (a topic that we have mostly avoided in this book; figure 1.4). When your dog gets its belly rubbed, it seems happy. But dogs don't rub one another's bellies, they don't pet each other's head, and they don't praise or scold one another with various tones of voice. These are clear examples of social communication that emerged in the close interaction between people and dogs, both through evolution and through

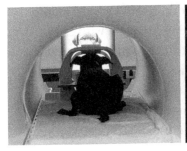
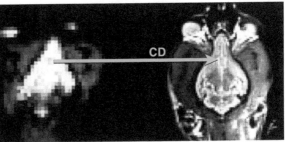

FIGURE 8.5. fMRI in dogs. *Left*: a dog inside an MRI scanner during an experiment in which the brain's response to other dog faces is investigated. *Right*: MRI data obtained. At the left are the functional MRI images, with statistically significant regions in yellow. At the right is a structural MRI showing the dog's brain inside the skull (the eyeballs are the two bright regions at the top). In the example shown, there was activation of the caudate (CD) when the dog saw a hand signal that indicated reward. Modified with permission from Dilks et al. 2015 (*left*) and Berns et al. 2012 (*right*).

learning. Figuring out what dog behaviors mean is thus very complex detective work: they may indicate emotional functions carried over from their wolf ancestry (for example, social submission), but they may also have generalized to indicate new functions (for example, play or instrumental behavior to get the owner to give food). We apply the names and concepts we use to describe our own emotions also anthropomorphically to our pets, but these are unlikely to be accurate in many cases. Figuring out what emotions dogs have can only be done through painstaking behavioral and neuroscience studies, most of which remain to be done.

Summary

- There are clear examples of links between emotions and brain regions in humans from lesion and electrical stimulation studies. However, these studies have a lot of confounds and poor experimental control, making it very difficult to understand the precise nature of that link.
- Two good examples are links between the amygdala and fear, and between the insula and disgust. Other brain regions, including the same ones as studied in animals (periaqueductal gray, hypothalamus), and including the prefrontal cortex, are involved as well.

- The future of fMRI studies of human emotions looks quite promising, even though the history of fMRI studies is fraught with experiments of poor quality. There has been rapid improvement in methods and a growing commitment to data sharing.

- It is possible to apply fMRI to some animals also. Future fMRI studies in monkeys and dogs may give us important insights into the homologies in emotion states between these animals and ourselves.

CHAPTER 9

The Neuroscience of Emotion in Humans

fMRI Studies of Emotion: The Logic and the Challenge

Having a tool like fMRI available is one thing, designing an experiment is quite another. Even aside from all the issues of extracting signal from noise, computing meaningful derived measures, and using statistical tools to draw reliable conclusions, there are challenges in how to design an experiment to manipulate emotion states. In studies in animals, we can present an animal with an ecologically meaningful stimulus, like the odor of a predator, and we can assay the emotion state through monitoring behavior, like freezing or running away, together with endocrine (hormonal) and autonomic (blood pressure, pupillary dilation) measures.

None of this is easy to do in humans; we generally cannot confront people with stimuli like predators, and even if we could, they would know that it is just an experiment and not respond in an ecologically valid fashion. We also generally cannot have people scream or run away, in good part for ethical reasons, but also because fMRI is exquisitely sensitive to motion (one of the reasons it is so difficult to do fMRI in awake animals). Even very small movements due to somebody's heart beating or breathing introduce substantial noise into the fMRI signal that require difficult corrections; having somebody scream would certainly ruin the measurement. On the other hand, it may well be that part of this movement noise could in itself be an informative signal—if you tremble, even if only slightly, that would show up in the fMRI signal, and, in the case of assaying a state of fear, actually carry information about the emotion state (but it would not be a neurobiological measure). As an aside, something like this has recently been found—you can identify individual people and predict many interesting individual differences (like IQ) just from their resting-state fMRI profile, but it turns out some of that information may come from the pattern in which people move in the scanner (due to breathing, scratching, and so forth). The movements create motion artifacts in the fMRI signal that can be as large, or larger, than the actual hemodynamic correlates of neuronal processing we want to measure.

The fundamental logic of an fMRI study is to experimentally introduce a contrast (or difference) in the quantity or quality of the variable of interest. So, you would want to contrast a condition in which there is an emotion state versus a condition in which there is no emotion state, and see where this difference reaches a statistical threshold in terms of activation in the brain. If you find a difference, you're in business; if you don't find a difference, you can't conclude anything since it may simply mean that you had inadequate statistical power to detect a difference. Many studies have a more parametric version of such a simple all-or-none contrast; they compare a condition in which there is a lot of an emotion versus one in which there is little, or have emotion vary in a parametrically graded way across trials that can be quantified with some function. For example, you could parametrically vary how much fear people experience, or how scary a stimulus looks, or how much people are required to think about fear, or even how much certain bodily signals (such as blood pressure or heart rate) change in response to an emotional stimulus.

A so-called event-related fMRI design would vary your parameters on each trial, but this may not be a good idea if you are studying emotion, since emotions are too slow (a trial typically lasts about 3 to 10 seconds). Instead, you might want to vary your parameter of interest along several levels in blocks, perhaps a minute or so long each. You would need to match the "persistence" feature of the particular emotion under study to the duration during which you are collecting the fMRI data (the length of your block).

Once you have such a parameterized fMRI study, you can then take the function that describes how the emotion changes as quantified by your parameter (measured from something other than the brain activity, such as verbal ratings or another behavioral measure), and ask where in the brain there is a corresponding change in activation that is fit by the same function (or that function multiplied by some justified transform). Roughly, you are interested in the following question: as you change something (measured) about the emotion (for example, report of intensity or valence), what is the corresponding change in the brain activity that you might pick up with fMRI?

It might seem relatively straightforward to decide what stimuli to use to induce an emotion state, but this is not at all so. Even in animal studies, inbred strains of mice may not respond to the odor of a predator the same way that wild mice would, and there is substantial variability across animals, in the same animal across testing sessions, and across laboratories trying to replicate the effects of innate emotion-inducing stimuli. The situation is worse in the case of humans; virtually *nobody* responds the same way to an emotion-inducing stimulus, and it is not even clear what an "ecologically valid" stimulus would be, exactly. A wild tiger? A tiger in a zoo? A movie or photo or cartoon of a tiger? The problem is that people *think* about the stimuli, the experiment, the expectations of the experimenter, their own expectations of the emotion, and so on, vastly complicating the brain activations produced.

The difficulty in setting up a clear and valid experiment is compounded by the fact that human emotion studies have often not focused on a small set of basic emotions, but instead have jumped right to investigating more complex emotions like shame and guilt and pride and Schadenfreude and love, and more. There are studies that have used words for emotions whose meanings neither the experimenter nor the subject fully understands. Moreover, the emotions investigated are typically quite subtle in intensity, and the studies require language, memory, and reasoning to conjure up the emotion, or a thought about an emotion. So, right at the outset, there is a major challenge with respect to construct validity (box 8.1): how do you know that you're studying an emotion, how do you know which emotion you are studying, and how do you know which aspect of emotion processing is reflected in your brain measure? While this point also applies to animal studies, it is even more pressing in the case of humans.

Lessons from Two Examples: Music and Faces

Let's take a look at this challenge with respect to two classes of stimuli that have been used in many fMRI studies on emotion. The first is music. When you play certain kinds of music to certain people, they will tell you that it causes them to feel strong emotions. Music, tone of voice (prosody), and other auditory stimuli can be quite effective in eliciting an emotion, although they are usually even more effective when combined with visual stimuli (such as in a movie with soundtrack, for example). When you ask subjects *what* emotions these are, specifically, it is much harder to get a clear answer (although you can steer them in the right direction if you play them music types with which they have specific cultural associations, and you give them a short list of emotion words from which to choose the best one). When you try to measure behaviors related to these emotions, it's even harder.

One study was particularly inventive in using a specific behavioral response; researchers played music to subjects that caused the subjects to report feeling shivers run down their spine (Blood and Zatorre 2001). This study in fact used positron emission tomography (PET), which in many ways was probably better suited at the time than fMRI (PET is much slower than fMRI, and so better matched to the time-course of an emotion; it can have better signal-to-noise; and it can be better at collecting signal from subcortical structures). The study found activation in a number of regions implicated in reward and emotion, including the amygdala, the orbitofrontal cortex, and the insula (cf. figure 8.1). One strength of this study's design was that it could use the very same sensory stimulus (the same piece of music) to make one person, but not another, feel shivers running down their spine; in this way, the brain activation that was correlated with feelings shivers run down your spine was not confounded with some aspect of the sensory stimulus (such as just louder music, or a particular tone or harmony). However, this came at the expense of a somewhat unusual measure of emotion—people's subjective ratings of shivers running down their spine. To increase the validity of this measure, the researchers also measured psychophysiological variables and found that when subjects reported feeling chills, they also had increased heart rates, changes in respiration, and changes in electromyography (EMG, a measure of producing facial expressions).

Music and other emotional auditory stimuli activate auditory cortices—which would process any kind of sound—but they also activate brain regions thought to be important for the experience of the emotion, such as the insula (figure 8.1). We already encountered the insula (for example, in relation to disgust) and it turns out to be activated by many emotional stimuli, and also by sensual touch and pain as well as by different tastes (sweet, bitter, and such). This makes some sense, since the insula contains an interoceptive sensory cortex that represents our perception of the state of our own body, as we already mentioned in chapter 8. Feeling something in your body, attending to something in your body, or just being sensitive to something in your body, all increase activation in the insula. Indeed, other studies have shown that the sensitivity with which people can detect changes in their own body, such as the rhythm of their own heartbeat, is correlated with activation of the insula. All these findings will remind you of the theory of emotion that William James once held, which we mentioned in earlier chapters. James thought that an emotional reaction in our body was in turn sensed by the brain and was the basis for our conscious feeling of the emotion. However, a major question is still left open by the study of music-induced chills. While the participants clearly felt something, was this an emotion? Which emotion was it? Which properties of emotions that we listed in figure 3.2 apply?

Studies that use stimuli like music or faces to probe emotion processing are confronted with a tricky question about what it is that people actually do in the experiment. In many cases people are shown scenes or faces, told stories, or hear music or voices, and are asked something about emotion. For instance, one of us conducted a study in the past in which we played music to people and asked them, "what emotion is this?" If you think about it, that is a very unclear question. It could be asking subjects to report on what the music is stereotypically intended to convey (it might sound like the soundtrack for a horror movie, for instance), what the composer felt or intended, what the musicians felt or intended, or what the listener thinks about or feels when hearing the music (which could be extremely idiosyncratic). All of these are rather different questions.

So music offers a mix of strengths and weaknesses for investigating emotion: it can certainly induce strong feelings in particular people, but it can also be unclear what components of emotion processing it engages, and unclear what specific emotions it can elicit. Depending on the task and the design of the imaging study, an experiment might reveal activation correlated with the experience of an emotion, or with thinking about a particular emotion category. Similar comments apply to movie clips, which are also commonly used in studies of emotion. Here, there are better normative data available in terms of the emotions that people subjectively rate themselves to feel when they watch the movie clips (for example, Gross and Levenson 1995). Still, there are major effects of familiarity, culture, and individual differences as well as the general complexity of a rich, time-varying stimulus. Thus, potent inducers of emotions in humans, like music or movies, in good part because of their very richness, also feature considerable variability. We don't know exactly what subjects are processing when presented with these stimuli, and there are likely big individual differences in all the emotion-related processing that they do (but we could in principle find out where in the brain these are generated). Finding stimuli to induce clear emotion states in humans in a uniform way is much, much trickier than one might have thought at first.

A simpler type of stimuli are those that are static and can be used to probe more time-locked, putatively stereotyped, emotional responses. By far the most common such stimuli used in human emotion studies have been facial expressions. This is largely because of influential earlier work by the psychologist Paul Ekman, whom you may remember from chapter 1. Ekman argued that human facial expressions can signal specific basic emotions in all cultures (Ekman 1994). So this seemed to be a particularly reliable type of stimulus, since everybody agreed on what emotion it showed. Unfortunately, there are problems here as well. First, Ekman's claim about the universal recognition of emotions from facial expressions has been hotly debated (Fridlund 1994; Gendron et al. 2014; Fernandez-Dols and Russell 2017), as we noted in box 3.5. Second, the way the stimuli are typically shown in fMRI studies (and behavioral studies, too) is not at all ecologically valid (they are single, posed, static pictures of faces on a computer screen without any background or other context). Third, just like with the music, it remains unclear what subjects are actually doing when they look at these facial expressions; they might be categorizing or naming the emotion shown in the face, since this is often required in the tasks given to subjects; they might be doing this by asking themselves what the person shown in the image is feeling (which, in reality, is not the emotion shown since they are typically emotional expressions posed by actors); or they might be asking themselves how viewing this facial expression makes them feel themselves (or empathically imagining how they would feel if they looked like the person shown in the picture). Probably, they are doing a bit of all of these. Certainly, facial expressions are quite a different kind of stimulus than are music or movies, and it would require a lot of further evidence to argue that they induce any emotion state at all, and if they do, it is a much less intense emotion (just looking at a facial expression of fear doesn't really make you afraid).

Two fundamental challenges apparent in emotion studies in humans are thus that the stimuli are generally not ecologically valid, and that it is very difficult to isolate specific components of emotion processing. We turn next to studies of human emotion that have investigated particular aspects of emotional processing, giving us an overview of the different kinds of processes, approaches to studying them, and a preliminary view of brain structures involved.

Attributing Emotions to Others

Many studies have focused on how we make attributions of emotion to other people, either from watching their behavior or finding out something about them or their circumstances. These studies thus focus on the last emotion property we had in our list in figure 3.2: social communication. The ability to attribute emotions to others depends on a network of brain structures that are involved in several different processes. At the front end, you first need to orient toward and attend to specific sensory cues, such as somebody smiling or frowning, laughing or crying. Once you have noticed these cues, they are processed further, often with additional contextual information, and they can then be categorized into an emotion category (among other categories): the other person is judged to be happy, or fearful, or sad, for instance (this is often imposed by the task subjects are asked to do). But processing does not stop there; you typically think more about why the person might be

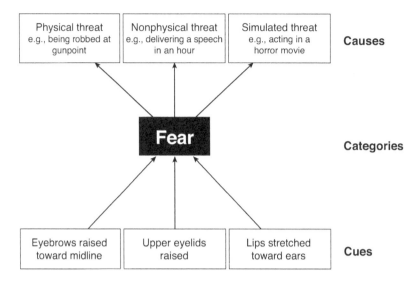

FIGURE 9.1. Attributing emotions and their causes. Although emotion attributions to other people often seem instantaneous, they consist of several stages. First, you have to detect salient cues; in this example, visual cues on a fearful face. Second, you infer an emotion category, together with other concepts (such as that the person may be about to run away). Third, you attribute a complex set of causes. In reality, all these three processing components interact. For instance, expectations about the highest attribution level (causes) can already direct our attention to look for particular cues to detect.

feeling the way that they do, about the causes and explanations for the emotion. (This is sometimes referred to as "theory of mind" or "mentalizing.") Depending on the nature of the initial information available, you might skip some of the first steps and go straight to the emotion categorization—for example, when you read a story about somebody, rather than actually observe or hear them. This multisequence picture of how you attribute emotions and their causes to other people is schematized in figure 9.1.

The brain systems that are associated with these different processing stages have been worked out in some detail through fMRI studies. Some of the key regions are schematized in figure 9.2. The amygdala is often activated in association with the rapid detection of salient cues—such as detecting wide eyes in a face that signals fear, consistent with data from studies of patients with amygdala lesions, who are impaired in their ability to attend to such cues (box 8.2). Working together with sensory cortices, the amygdala and other subcortical structures (such

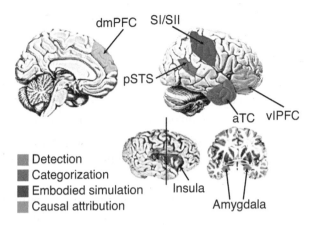

FIGURE 9.2. Brain regions involved in emotion attribution. There are many more structures involved than only the ones listed here, which are intended just to highlight some examples. According to some theories, the insula and other somatosensory-related cortices (SI/SII; purple) subserve emotion attribution by letting the observer feel the emotion that they see in the other person, possibly providing a richer account of how the emotion would feel through simulating it in the perceiver. dmPFC: dorsomedial prefrontal cortex. pSTS: posterior superior temporal sulcus. vlPFC: ventrolateral prefrontal cortex. aTC: anterior temporal cortex (also called temporal pole). Reproduced with permission from Spunt and Adolphs 2017a.

as the pulvinar nucleus of the thalamus) help us to orient toward, and to draw our attention to, those aspects of a stimulus that contain emotionally relevant information. Some of this initial detection may be quite coarse, and can be quite rapid.

The next step is more complicated, as we need to classify what we see into an emotion category. There are a number of brain regions involved in this step, such as regions that represent higher-level sensory information (for example, information about biological motion that is relevant to an emotion, processed in the superior temporal sulcus [pSTS]), regions from which conceptual and lexical knowledge about the emotion category can be retrieved (such as cortices in the anterior temporal lobe [aTC]), and regions involved in empathic responses to what we see (such as the insula, and related somatosensory cortices [SI, SII]). Finally, the most complex causal attribution to explain why the person whose face we are viewing is behaving the way that they are, and why they have the emotion we have just categorized, involves specific regions in the prefrontal cortex (vlPFC, dmPFC). One of these, dorsomedial prefrontal cortex (dmPFC), appears to be quite ubiquitously activated whenever we make causal inferences—it is activated not only when we infer the emotions of other people, but also when we infer emotions in our observations of animals (or even in ourselves). It would have been activated in Darwin's brain as he was making all of the observations of animal behavior that were the sources of his book, *The Expression of the Emotions in Man and Animals*. Damage to this brain region results in impairments in the ability to infer emotions in other people. The dmPFC also appears to be less activated in clinical populations who have difficulties in inferring emotions, such as people with autism.

All these studies bring us back to a question we already raised in the previous section (see also box 3.5). We know that people are attributing emotions, and we know something about the different stages of these processes. But what emotions are people attributing? What counts as a "normal" emotion attribution, and what lets us decide whether somebody's emotion attribution is correct or incorrect? It seems like simply taking the opinion of the experimenter, who is making his or her own emotion attribution about the stimuli used for the experiment, is not a very objective metric. Some recent studies have used representational similarity analysis (chapter 8) to address this question.

Imaging Emotion Concepts

Let's take a look at one such study, carried out by Rebecca Saxe and colleagues at MIT (this is the same person we saw together with her baby in the scanner in figure 1.1). Their study illustrates the application of similarity spaces, and of multivariate approaches (Skerry and Saxe 2015). The study made use of the representational similarity analysis that we described in the previous chapter and asked whether this could be used to understand how people categorize emotions—do they represent them in a 2-D space of valence and arousal, a popular psychological scheme we mentioned in chapter 3, do they map them into six "basic" emotions that psychologists like Paul Ekman have proposed (happiness, fear, anger, etc.), or do they assemble them from a larger set of properties, like psychological appraisal theories propose? (we discuss appraisal theories further in chapter 10). The researchers set out to answer these questions with the experiment schematized in figure 9.3.

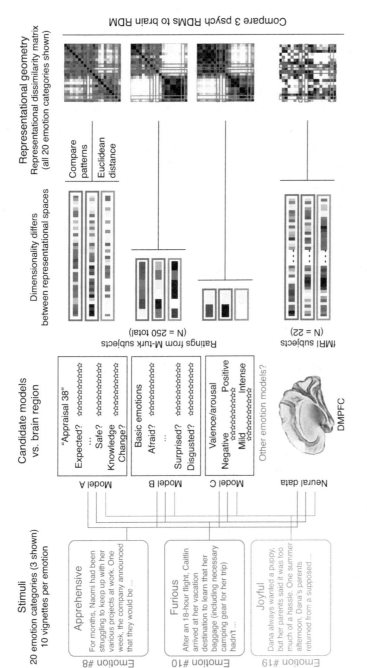

FIGURE 9.3. Representational similarity analysis of emotion concepts. The figure schematizes the flow of an experiment that used multivariate RSA to ask how people conceptualize emotions, from reading brief written scenarios (Skerry and Saxe 2015). See the text for details. Reprinted with permission from Dubois and Adolphs 2015.

This study was about how people attribute emotions to other people, and even more specifically, it was about how people attribute emotions to others from descriptions of situations in which people find themselves. So, this is a bit different from our previous examples, where we took the case of attributing emotions from actual observations of people (for example, their facial expressions). Once again, the study was not about actually having an emotion in any sense, but rather about how conceptual knowledge about emotions is represented. In the study, the participants in the scanner were presented with short vignettes (far left column) that described situations strongly associated with specific emotions. These vignettes had been carefully selected based on the ratings of emotions provided by an independent group of subjects who were first tested over the internet. There were several vignettes for each emotion, as indicated by the stack of stories shown in the left column, so that the researchers could look at the brain activation pattern that all stories for a particular emotion shared in common, rather than confounding the emotion with the particular content of any single story. The participants were not shown the words for the emotions, just the short vignettes. The pattern of brain activation elicited by reading the vignettes for each different emotion could then be subjected to representational similarity analysis (RSA) as we described in chapter 8 (this is the bottom row, illustrating the example of RSA as applied to a particular region in the brain, the dorsomedial prefrontal cortex, DMPFC).

So much for the brain data. The other type of data were psychological ratings, obtained for the same vignettes. Here, a different group of subjects were asked to produce ratings (not in the scanner, but again over the internet, using a platform called Mechanical Turk) that mapped the emotions into each of three different similarity spaces (coordinate systems). One had 38 dimensions (axes) describing many different features of emotional situations, including many that appraisal theories have proposed are in fact evaluated when a real emotion state is induced ("appraisal 38"), one had the six emotions proposed by Paul Ekman ("basic emotions"), and one had two ("valence/arousal"). The primary question now was: with which of these three psychological similarity spaces did the brain RSA correspond best? Note that the subjects in the scanner were only reading the vignettes, and that the psychological ratings all came from a separate group of subjects. So, when just lying in the scanner and thinking about the emotion associated with each vignette, how did the brain represent the different emotions? In terms of valence/arousal? In terms of basic emotions? Or in terms of 38 different appraisal dimensions?

There was some correspondence between the brain representational dissimilarity matrix (RDM) and each of the three psychological spaces (far right column), but the 38-dimensional one from appraisal theory fit the fMRI data best in this study, even when one accounted for the effect of merely having a larger number of dimensions. This best fit from the appraisal model may well have resulted from the nature of the stimuli: they forced subjects to appraise the situation along several dimensions, since that is how the stories were presented. It might well be that, if the subjects had been shown facial expressions of emotion instead, and we did the same kind of analysis on those data, then the six-dimensional or two-dimensional spaces would have fit best. But the overarching point is simply to illustrate how fMRI data, and psychological ratings, can be used to compare psychological and neural similarity spaces for emotion concepts. Note that in this study, once again, the dorsomedial prefrontal cortex was reliably activated in making attributions of emotions—a region of the brain we already saw in figure 9.2.

There are many other fMRI studies that have investigated how people attribute and think about emotions. One very important line of work remaining to be done is to characterize individual differences, and indeed cultural differences, in the way the brain represents emotions. All of the studies we previously described used educated, Western, subjects (typically, college students at a university in the United States). Different people make somewhat different judgments about the emotions another person is feeling, and different cultures even more so. Asian and Western cultures, for instance, look at different parts of the face in order to make judgments about the emotion that is expressed. Asians look more at the eyes and less at the mouth, a finding even reflected in the emoticons most commonly used in these different cultures (Caldara 2017). It would be very interesting to re-run the study above in a completely different culture. We would bet that there would be differences in how people from another culture interpret the vignettes, and in how they would rate the emotions that they think the people in the stories are feeling. This should be reflected in their brain activations. Behavioral studies of such differences are a large body of work in psychology, and have been for some time, but their neural basis has seen only rare investigation thus far.

These kinds of studies, then, are studies of how people *think about* emotions. They are studies of semantic knowledge, or of the concepts people have for emotions. They would also be closely related (though not identical) to the words that people have for describing emotions. This is an important line of work, but of course it is important to distinguish it from investigating emotion states as such—it really seems like a very different topic than the studies of emotion in animals we reviewed earlier. Presumably, many of these studies, certainly the one from figure 9.3, could not be done in nonhuman animals at all—but nonhuman animals surely have emotion states. A rat may not be able to think about, or have a concept (let alone a word) for "fear." But it can be in an emotion state of fear. There are many studies on the neural regions involved in emotion states in rats (for example, the amygdala and hypothalamus, structures we encountered in chapter 6), but there are no studies of how rats might attribute emotions to other rats or to people (although there are a handful of intriguing neuroscience studies suggesting that emotions can be socially communicated even in rats). It would be very difficult to know how to even obtain behavioral measures from animals to ensure that they are making emotion attributions. So, to a large extent, the neurobiological study of the attribution of emotions, or of concepts for emotions, is unique to humans and has no animal counterpart. (There is an ongoing debate about abilities for making some social attributions, often called "theory of mind," in other animals, especially primates—but this is likely a rather different ability than human emotion attribution, and next to nothing is known about its neural substrates in animals.)

There is an interesting twist to the topic of this section: attributing emotions and thinking about emotions is also precisely what the emotion scientist does! So, if we got a bunch of neuroscientists who do work on emotion together, we could put them in a scanner if we were interested in investigating how their brains make scientific attributions of emotion states from various kinds of data. This is not as silly as it sounds, actually. One could imagine contrasting two groups: take psychologists who only work on human emotion, and take basic neurobiologists who only do circuit-level animal studies. Present them with various kinds of stimuli on the basis of which one might infer emotion states (plots of experimental results, observations of animal and human behavior) and ask what kinds of attributions they make behaviorally, and what brain systems they engage when they make these attributions. Indeed, one could even do an interesting within-subject study: if you put the authors of this book in a scanner, you could ask us to look at animal behaviors while putting on our scientist hats and trying to evaluate whether these are emotions, or you could ask us to just think of these animals as our pets. Probably, we would make somewhat different inference in the two cases, and our brains should reflect the difference.

Feeling Emotions

Another large line of work in humans, that again has no proper animal counterpart, is the investigation of the conscious experience of emotions: what people feel when they are in an emotion state. This is a large and complex topic, since it invariably also involves attributions and concepts and words—the same processes we just reviewed in the previous section—only applied to oneself rather than to another person. Arguably, this is the core topic with which human studies of "emotion" have concerned themselves, explicitly or implicitly (although, again, most studies have not attempted to disentangle the experience of emotions from any other component of emotion processing).

The first important point to make is that, unlike for emotion concepts, or names for emotions (what we discussed in the previous section), which cannot be investigated in animals because they don't have them, experiences of emotion could, in principle, quite plausibly be investigated in animals. As we noted earlier (chapters 2 and 5) there is no reason to think that animals don't have conscious experiences of emotions—we just don't have a good dependent measure to detect them (there's no construct validity; box 8.1). That is, it seems unlikely that animals have concepts for emotions and think about emotions, but it seems quite plausible that they can feel them. If we grant that higher animals can feel anything, such as feeling pain, surely we should grant that they can also feel emotions, at least in principle.

We will discuss feelings a bit more in chapter 10. Here we only wish to make two points: a comment on the type of data typically used to study them, and an example. The data typically used to study feelings are verbal reports that people give. These are certainly an interesting source of data about emotions, but they are far from the only source of data, and they are problematic for several reasons. One reason is that it remains unclear on what exactly people are reporting when they are asked to report how they feel. People probably give ratings that are a mixture of their ability to judge their own feelings, their expectations of what the experimenter wants them to report, and their judgments of what people in general might feel or report. This is not to say that one cannot get highly reliable verbal ratings of feelings (for instance, by looking for consistency across multiple questions); there is also the advantage of experience sampling through verbal report in everyday life. But we believe that verbal ratings should be supplemented by other behavioral measures in order to quantify feelings, and we believe that they are not essential in order to measure feelings, so that feelings in principle could be studied with other measures in nonverbal animals (and nonverbal humans).

In figure 9.3 we saw a sophisticated application of multivariate fMRI to the investigation of how human subjects attribute emotions to other people. Multivoxel pattern analyses and RSA have been used also to investigate emotional feelings. In several studies, a variety of complex stimuli such as music or videos have been used in order to induce specific emotion categories, as we described in the beginning of this chapter, and multivoxel decoding was then used to build a model that described the patterns of activation that best distinguished among the different emotions. Using the kind of predictive framework we mentioned earlier (where only a portion of the data are used to build a model, whose predictions are then tested on the remaining set of data), classification models built from fMRI data can be used to predict what specific emotion a subject felt, from their pattern of brain activation observed.

In one ingenious study, for example, the researchers trained such a model to predict the brain patterns that would distinguish among a set of basic emotions, using the activity evoked by films and music. They then applied this model to predict what emotion people were spontaneously feeling, even if only subtly, while just lying in the scanner doing nothing (Kragel et al. 2016). The model predicted that people mostly felt neutral and a little surprised, and not particularly content, and this was then validated against the ratings that people actually give while they are lying in the scanner. This sounds pretty neat—the study used fMRI to read out, from people's spontaneous, resting brain activity, how they feel when there is no stimulus or task at all!

As you can probably surmise, it is unclear, however, what exactly this study was actually detecting. For instance, it may well be that people tend to think of particular things (have particular associations) when hearing a happy melody, and they also think of these kinds of things when they feel happy while doing nothing and lying in the scanner. Perhaps they often imagine being by a clear mountain lake, or imagine blue sky and sunshine, when they feel happy, for example. If this were the case, the fMRI study just described might have trained its multivariate classifier to pick up representations of lakes and blue skies and sunshine in the brain, among many other things—that is, neural activations that indeed are predictive of whether somebody is feeling happy, but are not constitutive (are not part of the direct causal mechanism) of feeling happy. All kinds of things might be able to predict whether you are happy rather than sad, but they need not tell me anything about the mechanism whereby the brain generates these emotions. This is just like the distinction between causation and correlation we discussed in chapter 4; there, we noted that the speedometer in your car is a very reliable predictor of your car's speed—but has nothing to do with the causal mechanism that makes your car move.

Perhaps in support of this speculation, the study we mentioned above (Kragel et al. 2016) found that the multivoxel pattern that their predictive model used was in fact distributed all over the brain, and not in any specific structures or system. This then highlights one of the challenges with the powerful multivoxel fMRI methods that are now being applied to the neuroscientific study of emotion; they can only show you which patterns reliably *predict* aspects of emotion, but they cannot unambiguously show you the actual neural *mechanisms* of emotion. They may show you neural *correlates*, but they fail to elucidate neural *causes*. In a sense, they may simply be a brain version of what we already have: we can reliably predict if somebody is feeling sad or happy or afraid just from observing them. Now we can also observe their entire brain, and the complex patterns we find there can do the same thing. That may be a predictive marker for an emotion, just like the behavior is, but it's not yet an account of the mechanism. It is possible that newer approaches for inferring causal structure from fMRI data, such as those we reviewed briefly in box 8.3, can give us models of causal mechanisms.

Another shortcoming with many studies of emotions in humans, which we already noted, is that the emotions are often rather weak versions of what one might study in animals. However, there are some studies that have in fact tried creatively to induce strong feelings of emotions that have some validity. One of the first such studies used PET (positron emission tomography). This study was led by Antonio Damasio at the University of Iowa, the neuroscientist we already encountered earlier for his work on patients with lesions to the prefrontal cortex. His team wanted to investigate the neural correlates of emotional experiences (Damasio et al. 2000). Like with the music study we mentioned earlier, they were acutely aware of the fact that different people might have quite different emotional experiences if you show them a movie or have them listen to music. To get rid of the problem of specific stimuli whose processing could confound the interpretation of the results, and to get emotional experiences that were as valid and strong as possible across every subject, they creatively chose a different approach.

The researchers asked subjects to think about autobiographical experiences that had evoked really strong experiences of emotions in the actual lives of their human subjects. Each participant was first asked to think carefully about episodes in their lives in which they felt strong instances of one of four emotions, happiness, sadness, fear, and anger. Once they had a clear image and recollection of this episode, this was then selected as a stimulus that subjects could be asked to recall in vivid detail in the scanner. The imaging experiment then contrasted blocks in the PET session in which a subject recalled and reexperienced one of these four emotions (from their personal memory episode), with a neutral episode (which also involved memory and imagery, but not the strong emotion). Since PET has very slow temporal resolution, the results reflect the average activation produced over the entire time window that subjects were reexperiencing the emotion.

There were strong activations and de-activations across the brain in this study. These included many brain regions, both cortical and subcortical, including ones that the authors had hypothesized in the study (regions involved in regulating, or in representing, body states that were hypothesized to be involved in the conscious experience of emotion). Once again, the insula and other structures related to representing one's own body state showed up.

This study was able to evoke strong experiences of emotions (validated from ratings given by the subjects in the study), and it also averaged out the specific details of the recalled episodes (just like the music study we mentioned earlier averaged out the specific sensory properties of music pieces since different subjects found different pieces of music to send shivers down their spine). Contrasting experiences of particular emotions with experiences of relatively neutral episodes that the subjects also recalled further helped to remove activation that would be associated with the memory recollection, with attention, and with aspects of maintaining the images in mind during the experiment. Nonetheless, this study was of course not able to dissociate the conscious experience of the emotion entirely from either the emotion state or the emotion concepts. Subjects likely induced actual emotion states in themselves through their imagery, and may have been thinking about and categorizing the emotion that they felt. With the focus on attending to the conscious experience, and with the induction not through an actual stimulus but through autobiographical recall, it is plausible that the study provided a relative emphasis on the conscious experience of emotion, over the actual emotion state or thinking about the emotion concepts.

This experiment thus highlights a strategy that researchers conducting neuroimaging studies in humans will also have to think about in the future. It is likely going to be impossible to completely separate emotion states from conscious experiences from emotions, or from having people think about emotion concepts. But it may still be possible to emphasize one of these three components relative to the other two by having a task that requires people to attend differentially. If a larger number of such carefully designed studies eventually accumulate data over time, we should be in a position to examine differences in the neural correlates of these three aspects of emotion with more statistical confidence.

Central Emotion States

We end this chapter with a last section on actual emotion states. These are the aspect of emotion investigated in animals (the topic of chapters 5–7), the core topic of this book, and the aspect of emotion we would like to have clear studies on in humans as well. Surprisingly, this is the least investigated aspect of emotion in human studies. We know something about the neural correlates of words and concepts for emotions, and we know something about the neural correlates of conscious feelings of emotions (and both of these tend to be confounded with one another in most studies). We know very little about the neural correlates—let alone mechanisms—of actual emotion states, in part because clear strong emotion states are in fact rarely induced in neuroimaging studies.

Nonetheless, there have been a few studies that have attempted to induce strong, specific, emotion states in human fMRI that are closer to the kinds of strong, stimulus-linked and behavior-linked emotion states induced in animal studies. One such study was carried out by our Caltech colleague Dean Mobbs (Mobbs et al. 2010). While participants lay in the scanner, the researchers placed a tarantula next to their feet (which stick out as you lie on your back in the scanner). There was a slight bit of deception involved in the experiment, but this is what the subjects believed—and there was a real live tarantula used in the experiment! (figure 9.4).

The researchers were interested in how the brain responded to the imminence of this fear-inducing stimulus: what happened as the box containing the spider got closer and closer to the subject's foot? They

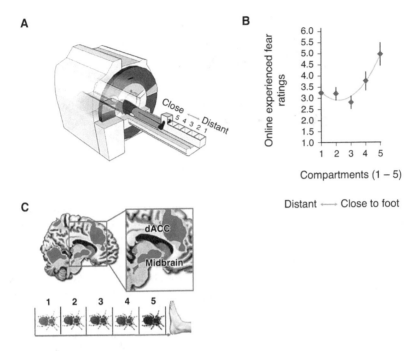

FIGURE 9.4. Inducing strong emotion states in an fMRI study with realistic stimuli. In this study, subjects were put into an fMRI scanner (A) while they saw that their foot was in a compartment. A tray of adjacent compartments was arranged such that some were closer or farther away from the foot, as shown in the figure. The experimenter then placed a tarantula into one of the compartments and measured the brain activation produced using fMRI. (B) As one would expect, ratings of expected and actually experienced fear increased the closer the tarantula got to the foot, as evidenced from numerical ratings that subjects provided with a button box inside the scanner. (C) The closer the tarantula got to the foot, the greater was activation in a network of brain structures. Notably, these brain structures included regions that have been shown to encode the dimension of threat imminence in other studies, such as the dorsal anterior cingulate cortex (dACC, also called anterior midcingulate cortex) as well as the amygdala and bed nucleus of the stria terminalis (not shown). Reproduced with permission from Mobbs et al. 2010.

found fairly specific brain activation in regions around the amygdala and bed nucleus of the stria terminalis, regions strongly implicated in fear and anxiety also from animal studies (chapter 6). Activation in some regions parametrically tracked the proximity of the tarantula (closer elicited greater activation in the midbrain and dorsal anterior cingulate cortex, and the magnitude of the activation correlated with the subjective rating of the intensity of fear that subjects experienced). Other regions tracked more complicated metrics—for instance, the bed nucleus

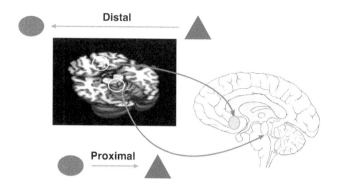

FIGURE 9.5. Threat imminence activates specific brain regions. In this f MRI study, subjects played a computer game in which they had to try to escape an artificial predator. The predator was schematized by a red circle in the game, and the subject showed up as a blue triangle in the game. As the predator roamed around the game board, the subject had to move a joystick to try to escape from the predator. When the predator caught the subject in the computer game, this had a real consequence: they got a painful electric shock. This study thus inventively combined the quantitative aspect of measuring threat imminence (parameterized as distance in the game) with the validity of a measured behavior and a real outcome (shock), all under the constraints of what one can do in f MRI (such as not actually moving). Whereas more distal threat activated the prefrontal cortex (orange in the brain schematic on the right), thought to implement planning and monitoring, more proximal threat activated the periaqueductal gray (green in the brain schematic on the right), thought to implement immediate defense reactions. Reproduced with permission from Mobbs et al. 2007.

of the stria terminalis tracked not the mere proximity of the tarantula, but whether (at the same distance) the tarantula was moving closer or moving away from the foot. Studies like this can begin to investigate how particular properties of a threat, such as its imminence, might be represented in the brain and used to trigger particular sets of adaptive behaviors, such as those we saw schematized in figure 2.4. Related studies nowadays are actively exploring the use of virtual reality to substitute for real spiders, opening up a large new domain for presenting realistic-looking stimuli without needing to have jars with tarantulas on hand (a boon also to those researchers who are arachnophobic).

In another study by some of the same researchers, electric shocks were used together with an escape game that subjects played in the scanner (Mobbs et al. 2007). This may not seem as ecologically valid as having a live tarantula, but it is still a perfectly valid paradigm, since there were real aversive consequences (real electric shock if a virtual predator caught you in the game), and since playing games to obtain rewards or avoid punishments is something humans commonly do in the real world. In fact, this study has better validity than the tarantula study because we have actual behavior from the subjects as they played the game—we can directly tie their emotion state to the avoidance behaviors that they show in the game. In this study, the researchers found that the imminence of threat from the electric shock resulted in a switch in the activation of brain systems, from ones prominently involving the prefrontal cortex (correlated more with anticipation and planning related to the threat) to ones prominently involving the periaqueductal gray matter (correlated more with immediate reaction to threat) (figure 9.5). Once more, this study featured a design that captured the ecological dimension of threat imminence, and related it to emotion states in the brain (see box 2.3).

Dissociating Emotion States from Concepts and Experience

Studies of this sort would seem like they could begin to show us the neural correlates of emotion states in humans—and indeed some of the brain structures that would have been hypothesized on the basis of animal studies are coming into play. On the other hand, just as with our concerns about isolating the conscious experience of emotion, here, too, there will be contamination with other emotion components.

Of course, there will always be contamination with many other components in a single person on a single trial, since the brain needs to perceive the stimuli, remember any task instructions, and attend to the task; so, perceptual, memory, and attention-related systems in the brain will be engaged as well. Emotion states cannot occur in isolation. The logic of an fMRI study is to experimentally manipulate just the emotion. The perception, memory, and attention would remain fairly constant across trials, but if we arrange it so, the intensity, or type, of the emotion would vary in a way that we could capture with a parameter in a model, allowing us to investigate how brain activation changes when the emotion state changes across multiple trials.

The real concern about confounds is this: is there anything that might be *systematically correlated* with emotion, so that, even across trials and subjects, whenever an emotion state varies, so do these additional factors? If that were the case, it would make it impossible for us to draw conclusions just about the emotion state, since such covarying factors could account for the brain activations that we observe in the study.

This is a difficult question to answer, because most of our independent (non-fMRI) measures of an emotion state in humans are indeed confounded with other factors. Notably, behavior is usually an indicator not only of the emotion state but also of the conscious experience of the emotion (but see box 10.1). Similarly, subjective ratings seem to track the conscious experience of the emotion as well as the emotion state—and indeed the concept of the emotion (since people need to put their feelings into concepts and words). It seems likely that whenever we induce a strong emotion state in humans, they consciously experience it, and so they begin to think about it—their brain begins to represent all the memories, associations, and words that come with that particular emotion. To put it bluntly: suppose there were a brain region for representing just the word "fear." Well, every time you actually induce a state of fear in subjects, they will consciously know they are in a state of fear, and they will probably automatically think of the word "fear" and activate this region, more or less so the more they think about fear, which could be correlated with the intensity of the fear state. So your fMRI study would find that this region, which is just involved in lexical representations, tracks the intensity of the fear state and is confounded with the actual neural mechanisms for fear. Again, much of the problem stems from inferring causation from correlation (see chapter 4).

It is notable that the similarity structures of different aspects of emotion are remarkably similar. If you ask people to rate the similarity of facial expressions of emotions, of words for emotions, of emotional situations, or of their own emotional experiences, these all show the same similarity structure. Moreover, so do the corresponding brain activations. Work by the Finnish neuroscientist Lauri Nummenmaa has documented this correspondence (figure 9.6). There is evidently a close relationship between the similarity space in which we can map emotion states, emotion experiences, and emotion concepts. Perhaps this is unsurprising, since in the end we are using the concepts to talk about all of these.

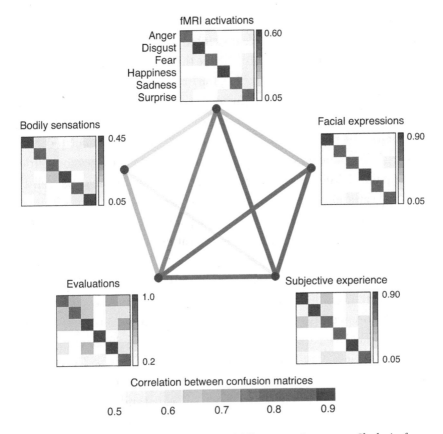

FIGURE 9.6. Similarities between the structure of different emotion aspects. Clockwise from top: similarity matrices derived from a pattern analysis of BOLD-fMRI activations during emotion experiences; human observers' ratings of facial emotions; direct pairwise similarity ratings of people's subjective experiences; Euclidian similarity of intensity profiles of discrete emotion ratings for short narratives; and a linear discriminant analysis of bodily maps of emotions (the same as shown in figure 2.2). The lines in the middle of the figure indicate the Spearman correlation between the different similarity matrices (big scale bar at the bottom). All matrices have the same axes, and the x-axis is the same as the y-axis. Modified from Nummenmaa and Saarimäki 2017 with permission from Lauri Nummenmaa.

Our ability to disentangle the different aspects of emotion is not as impossible as you might think, for at least three reasons. For one, we can in fact, at least to some extent, de-correlate or minimize what people think about or consciously experience when they are in a state of fear. Although emotion states and conscious experiences of emotions are strongly correlated, they are not perfectly correlated. A way to disentangle them using fMRI has been explored in the case of pain. In a recent study by Tor Wager and colleagues at the University of Colorado (Woo et al. 2017), subjects were presented with varying degrees of pain (stimuli with a hot temperature). Three things were measured: the objective temperature of the stimulus, the brain activation produced, and the subjective ratings of pain given by the subjects. The pain ratings were highly correlated with the temperature: the hotter the stimulus, the higher the pain rating. But they were not perfectly correlated—on any trial, the same temperature might be felt to be a little less or a little more painful. The researchers took this residual variability, as well as residual variability from brain activations in regions previously known to encode the nociceptive properties of pain, and asked what might be left over. What residual variance in the ratings might be explained by activation in other brain regions?

A sensitive multivariate analysis was used, training a machine learning algorithm on the voxelwise pattern of activation that was left over after the nociceptive signal had been regressed out. In a sense, this corresponded to all other cognitive activity, not directly correlated with the stimulus intensity, but correlating with the pain ratings. They indeed found that there was a network of other brain regions that predicted this residualized feeling of pain: the ventromedial and dorsomedial prefrontal cortex, and insula, among others. They cautiously called this the "stimulus intensity independent pain signature," but it might be (in part) a measure of the conscious experience of the pain, independent of its discriminative nociceptive properties. Taking a similar approach to the study of emotion could help us to separate the neural substrates of the emotion state from the neural substrates of the conscious experience of emotion. Of course, there are still caveats about this approach; for instance, if I have a tremor in my hand as I give the ratings, the residual variance in pain ratings would not reflect conscious experience of pain, but motor tremor. Nonetheless, the set of brain regions found may give us additional clues (motor cortex did not show up in the study, so tremor is not a likely explanation).

A second reason that it should be possible to tease apart emotion states, emotion experiences, and thinking about emotion concepts goes back to the study by Damasio in which brain activation to autobiographically recollected emotional experiences was investigated using PET. That study, like the others we have described here, likely imaged not only emotion experiences but also emotion states and emotion concepts. But the specific experimental protocol probably emphasized the conscious experience of the emotion over the other aspects of an emotion. By contrast, the experiment shown in figure 9.5 probably emphasized the emotion state over other aspects of the emotion. And by further contrast, the experiment shown in figure 9.3 probably emphasized emotion concepts over other aspects of the emotion. So, even though *individual* experiments may always involve multiple components of emotions, they do so differentially. If the experiments are carefully described, and if the data from them are made available for further meta-analytic studies, then we should be able to make comparisons across collections of these neuroimaging studies. Some studies will emphasize emotion experiences, some will emphasize emotion states, and some will emphasize emotion concepts. With care, and with a large enough number of such studies, it should indeed be possible to separate those neural activations that are the unique correlates of each of these three aspects of an emotion. This may well be the way of the future in neuroimaging studies on emotion, but it requires that different studies use comparable standards for stimuli, tasks, subject selection, and fMRI methods; that they are very thorough in describing their methods; and that they then openly share all the data so that it can be further aggregated and analyzed.

There is a third reason to be optimistic about our ability to distinguish different aspects of emotion processing in the brain. This brings us back to the caveats with which we already began chapter 8: in the end, it is clear that we cannot investigate emotion using only fMRI and only humans. We need to supplement it with methods other than fMRI in humans, and with experiments in animals. At a minimum, investigations in animal species can tell us where to look; there may be changes in many parts of the brain for which it is difficult to unambiguously tell whether this region is contributing to processing the emotion, or merely something caused by the emotion. But there may be other parts of the brain for which we have an a priori hypothesis, because we have strong evidence from other methods and other species, ideally evidence of a causal sort. If we have strong causal evidence that certain parts of the hypothalamus, or of the periaqueductal gray, are part of the mechanism for an emotion state across mice, rats, and monkeys, then finding such activations in humans would reassure us that we are inducing an emotion state (although not prove it, since this indeed uses reverse inference). Conversely, if we find some activations in a human study that have no homologue in any animal studies, this may reflect conceptual aspects that are not present in the animal studies. In the end, it is only such a comprehensive science that can uncover the neurobiological basis of emotions. If we begin to put together some of the causal methods from chapters 5–7 with techniques like fMRI discussed in this chapter and the previous one, it is possible to conceive of future experiments that could begin to provide much more definitive answers. We close this book by sketching some of these future experiments, in chapter 11. But before we end the book, we spend a little more time discussing some other views on emotion, and a little more time discussing the aspect of emotion most prominent in the human literature: feelings.

Summary

- fMRI studies require careful comparison between well-matched conditions, and impose severe constraints, such as requiring the participants not to move while lying horizontally in the fMRI scanner. This raises questions about the ecological validity of fMRI studies of emotion.
- Attributing emotions to other people involves a network of primarily cortical regions, such as the dorsomedial prefrontal cortex.
- The conscious experience of emotions activates structures that represent states of our own body, such as the insula.
- Central emotion states, uncommonly studied in humans to date, involve ventromedial prefrontal cortex, periaqueductal gray, and amygdala.
- While it is very difficult to disentangle different aspects of emotion processing, such as actually having an emotion, consciously experiencing the emotion, and thinking about the emotion, this can be done if we compare results across many types of studies. Notably, we should compare studies of emotions in humans and in animals.

PART III

Open Questions

CHAPTER 10

Theories of Emotions and Feelings

There are many theories of emotion experience (see Shackman et al. 2018). Many of these theories have their origin in psychology rather than neuroscience. Most of them focus on feelings, the conscious experience of emotions. Although this book has not been about the conscious experience of emotion, but about the neuroscience of emotion states, we provide an overview survey of other theories of emotion in this chapter. We do comment on how these theories relate to our own framework, but it is important to note that we ourselves have not proposed any kind of detailed emotion theory.

Indeed, we generally feel that it would be premature to have very detailed theories of emotion, at least insofar as the neurobiology is concerned, because the research required to underpin such theories is just beginning. It is also likely to be the case that different aspects of emotion might require different theories to do them justice, so that while an overall framework for thinking about emotion, and an agreed upon terminology, are both essential, a global "theory of emotion" that covers everything may not be that useful for actually formulating hypotheses. As biologists, we have similar opinions about global theories of cognition: brains are the product of millions of years of complex and often idiosyncratic evolution, not a single grand design. It might be most prudent to investigate how they function more locally, in specific systems and for specific topics, and to then abstract fundamental principles that are more descriptive than full-fledged "theories" of how the brain works. This is also the spirit in which we wrote this book—to sketch a path for scientists to investigate an incredibly complicated topic, not to offer a "theory" that purports to know how brains are built and how they work.

Notwithstanding these quick cautions about the scope and, in our view, the limits of what theories of emotions can offer at this stage of our knowledge of the mind and the brain, we will briefly survey some of the theories of emotion in the modern literature and contrast them with our framework. This will be a useful counterpoint to all of the previous chapters and will give us an idea of the range of hypotheses out there. At the outset, it is critical to note that apparent disagreements between our framework and other theories may often stem not so much from actual disagreements, but rather from differences in topic or terminology. For instance, many debates in emotion theory revolve not so much around how we should think about "emotion" but rather about specific emotions, like fear, anger, and so forth. That is, many of the debates are about the taxonomy of emotion, a topic we have mostly avoided in this book, rather than about emotion per se (figure 1.4).

The Structure of Affect

The layperson and the psychologist alike tend to focus on the subjective conscious experience of emotion, what psychologists often refer to as "affect" or "feelings." Feelings of course refer to a much broader range of conscious experiences than just experiences of emotions (I can feel tired, or thirsty, or itchy, for instance). The term *feelings* emphasizes conscious experiences of something happening in one's own body, and these kinds of interoceptive conscious experiences are often distinguished from other perceptual experiences, or from "cognitive" experiences (what we might typically call *thinking*). Since all of these phenomena commonly co-occur when one has emotional experiences, it is helpful to provide some classification scheme.

At the most basic level, you can be conscious (for example, a typical healthy, awake adult human) or unconscious (for example, a patient in a coma). If you are conscious, then the next question is what it is that you are conscious of; that is, what is the content of the conscious experience. We think there are three broad categories of conscious content: somatic (about things happening in your body), perceptual (about things happening in the outside world), and cognitive (about things happening in your mind). Some theories argue that everything is in your mind, or in your body, but for now we just take these broad categories as a starting point for describing your conscious experiences.

Our rough view is that a conscious experience of an emotion requires the following. First, you need to be conscious (not in a coma); this applies to humans and animals. Second, you need to have some somatic content to your conscious experience (how the emotion feels in your body); this also applies to humans and animals. Third, you need to have some cognitive content to your conscious experience (for example, a motivational component, like the feeling that you need to run away from a threat); this also applies to humans and animals. Note that humans additionally have the metacognitive knowledge of all these components (if they reflect on them); we know we are not in a coma, that we are feeling our body, and that certain things are going on in our minds.

A typical human conscious experience of an emotion then requires (a) that you are awake and conscious (the level of consciousness), and (b) that you are conscious of several things, such as feelings in your body, an urge to behave in certain ways, and some beliefs about yourself. One likely difference between humans and animals is that many instances of human feelings are caused by thinking about what might happen in the future, or reminiscing about what happened in the past, rather than about current stimuli. This is reflected in two differences between people and animals. For one, the perceptual content (for example, seeing a tiger), we believe, is a necessary component of the conscious experience of an emotion in animals, at least at the point that the emotion is induced. That is, animals require perception of some situation in order to trigger an emotion. Humans may or may not have an actual situation trigger an emotion—they can also induce emotions just by thinking about something or imagining something. It is likely that this additional flexibility for how emotions can be induced in humans also contributes to the increased risk that something can go awry in how emotions are induced, as occurs in psychiatric illnesses like depression and anxiety (see box 2.2). Another difference is in terms of cognitive content, where people often relate emotions to themselves or to beliefs that they hold about the emotion.

While experiences of human emotions are clearly very rich in their contents, most psychological theories of emotion propose that conscious experiences of emotions exhibit a simpler, underlying core dimensionality of valence and arousal, often referred to as "core affect" (Feldman Barrett et al. 2007). Valence, in particular, seems to be a necessary core ingredient of how emotions are experienced—they have to vary on a dimension of pleasantness/unpleasantness. Note that valence is also an emotion property we listed back in figure 3.2. There, we already noted that "valence" in psychology is an essential dimension of the structure of the conscious experience of emotion, measured by having people rate the experienced pleasantness or unpleasantness of how they feel. However, this sense of "valence" is distinct from the sense of valence that we used in figure 3.2, where we intended it to refer more broadly to an aspect of the emotion state—namely, the aspect that specified whether the emotion state is directed toward escaping something harmful or approaching something beneficial. That is, the notion of "valence" with respect to an emotion state (as in figure 3.2) refers to a motivational, behavioral, and functional aspect of approach/withdrawal, whereas the notion of "valence" typically used in psychological theories of emotion experiences refers to a dimension of the conscious experience that people can rate. These two senses of "valence" are of course closely related and usually highly correlated—but they are distinct nonetheless. It is an important open question whether these two senses of "valence" might be subserved by distinct neural mechanisms.

For the psychological sense of "valence" as the dimension of experienced pleasantness or unpleasantness in our feelings, then, essentially every daily experience has this kind of valence. We can ask at any time how pleasant or unpleasant somebody feels at this moment, even when there is no emotion. This broad assessment of an experienced valence dimension is shown in figure 10.1.

Most psychological theories of emotion experience further assume that while arousal and valence are necessary core ingredients of an emotion, they are not by themselves sufficient. One reason is that they are too ubiquitous. For instance, arousal can vary without a change in emotion state when I run slow or fast (see chapter 5). Much the same is true of valence. A smell or touch or taste can have a strong valence without necessarily inducing any emotion state. So "core affect," the underlying arousal/valence dimension of emotional experience proposed by many psychological theories, is taken to provide a kind of scaffolding, a "core," to which additional ingredients typically need to be added to constitute an emotion experience.

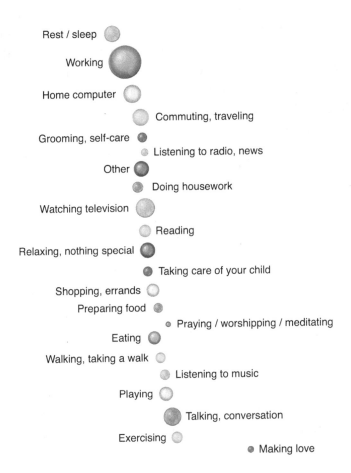

FIGURE 10.1. Feelings in everyday life. Data from experience sampling throughout people's daily lives are represented in terms of the activity (labeled bubbles), how often it occurred (the size of the bubble; bigger bubbles indicate more frequent activities), and how pleasant it felt (least pleasant on the left, most pleasant on the right). Modified with permission from Killingsworth and Gilbert 2010.

Most theories of emotion experience thus have a kind of layered architecture. They begin with what is assumed to be a necessary bedrock for experiences of emotion (such as valence/arousal, a sense of self, or interoception, depending on the theory), and then add more elaborate contents to the conscious experience (such as awareness of body states, awareness of situations and stimuli, or metacognitive self-awareness). Some of the lower layers in such architectures are thought to be common to humans and animals, and some of the higher layers are thought to be unique to humans. Some theories, for instance, propose self-reflective representations, or working memory, or language, as a uniquely human ingredient that gives conscious experiences of emotions their particular flavor seen in humans. These views make some predictions that could be tested with neuroscience studies: we might find reliable neural signatures of valence in human studies of emotion, but find more variability in how other aspects of emotion experiences are represented in the brain.

Theories of Feelings

Interoceptive Theories

It should go without saying that our survey of other people's theories of emotion in this chapter is of course our own interpretation of their theories; any errors are ours. Most psychological theories of emotion are about emotion experience but do not provide mechanisms whereby the experience is generated. Some exceptions come from two neuroscientists, Bud Craig and Antonio Damasio, who have proposed theories of feelings that feature a layered architecture grounded in interoception (the brain's detection of the body's internal signals), with quite specific mechanisms gleaned from neuroscience studies. These theories stress the role that representations of the body play, not only in emotion experiences but also globally in feelings and even in the level of consciousness. These theories share some commonality with William James's idea that perception of one's body states leads to emotional feelings. Craig and Damasio both emphasize the insula (figure 8.1), a sensory cortex that maps input from the body, including pain signals as well as autonomic and visceral information. According to Craig, there is a progressively more elaborated representation of the body, and indeed of the person as a whole, as one goes from the posterior to the anterior insula (Craig 2002). The insula, in Craig's theory, thus represents, in a comprehensive way, how one feels, including most or all of the components of a conscious experience of an emotion that we listed at the beginning of this chapter.

There are multiple pathways from the body to the brain that can implement the route William James had hypothesized. One route that has received considerable attention in emotion research involves afferents from the autonomic nervous system, which carry bodily sensory information from all the viscera to the brain, and eventually relay information to structures such as the anterior cingulate cortex and the insula. Whereas Craig suggests that these pathways to cortical regions may be sufficient for our subjective experience of how we feel, Damasio stresses that feelings arise from body-state regulation that is more distributed throughout the nervous system, and that importantly includes brainstem and midbrain as well as cortex (Damasio 2003; Damasio and Carvalho 2013). For instance, one study tested a patient whose anterior cingulate cortex and insula were lesioned, but whose primary somatosensory cortex was intact. The patient was able to detect changes in his own heartbeat normally, which the study took as evidence of intact interoception and at least partial evidence for the claim that feelings do not require the anterior cingulate cortex and insula (Khalsa et al. 2010). Whether the structures left intact in the patient's brain would be sufficient to yield a normal subjective experience of emotion, however, still remains unclear.

Damasio's theory of feelings points out that there are many representations of the body, and many loops for homeostatic regulation of internal organs (cortical and subcortical). These all contribute to conscious experience in general, and they all contribute to the conscious experience of emotion as well. Damasio has proposed that a primordial representation of the organism's body is implemented in these multiple levels and generates not only feelings but also a basic sense of self. According to this theory, this provides the grounding for all of conscious experience—a prerequisite representation of the self, to which other representations (for example, of objects and events in the world) can then be referred. When we say "I am feeling afraid of the tiger," Damasio's theory would deconstruct this as a relation between a representation of the self ("I am afraid") that includes a comprehensive representation of the state of the body, relative to its normal homeostatic state, and a representation of the tiger that one is seeing. In this view, feelings and conscious experience are two sides of the same coin, and it is impossible to have conscious experiences if you do not have feelings and interoception (Damasio 1999).

In relation to what we have written in this book, these theories of the conscious experience of emotion offer a straightforward and attractive addition. Whereas the emotion state may be investigated in specific circuits, such as those involving hypothalamus, amygdala, and other subcortical structures (chapter 6), the conscious experience of those states in humans (and, possibly, in animals) would additionally require a number of further ingredients, such as representations of the self and representations of the state of the body.

Appraisal Theory

There are also influential psychological theories of emotion experience that show layers of sorts. One class of theories is appraisal theory, which has a long history from early writers like Magda Arnold and Richard Lazarus (Lazarus 1991) to more modern psychologists like Klaus Scherer (Scherer 1984; 2009). The idea behind appraisal theories is that there is a sequence of evaluations of a stimulus or situation that progressively makes more complex decisions. Aspects of this hark back to our discussion of drift-diffusion models: there is accumulation of information over time, leading to a number of distinct decisions. We also briefly encountered appraisal theory in the previous chapter, where we described the experiment investigating the conceptual similarity spaces by which people represent emotional situations (figure 9.3). There, we saw that the components proposed by appraisal theory actually seem to fit fairly well with how people attribute emotions to situations, from their neuroimaging data. Perhaps this is unsurprising; it just shows that the psychologists who came up with appraisal theory were reasonably in touch with how they themselves attributed emotions intuitively.

Although appraisal theories are generally presented as being about the conscious experiences of emotions, they have a strong functional character and can easily be applied to emotion states, without needing necessarily to say anything about feelings. The work of Richard Lazarus (Lazarus 1991), for instance, proposes several specific functional modules, fleshed out as so-called "core relational themes," that describe the adaptive challenges an organism would face, and the functional steps it needs to take in order to cope with that challenge. While heavily focused on humans and on the human experience of emotion, appraisal theory is ultimately a functionalist theory.

The psychologist Klaus Scherer has provided a very detailed theory of how situations are appraised in light of their adaptive significance, and in light of the ability of the person to cope with whatever challenges the situation offers. According to Scherer's theory, an emotion unfolds over milliseconds to seconds as we progress from detecting the relevance of an event, to evaluating its implication, our ability to cope with it, and its normative significance. So, these correspond to layers of more and more elaborated processing in time, as further types of information are considered. They involve multiple aspects of an emotion that all need to be coordinated, ranging from autonomic response to motor expression and feelings (Scherer 2009).

The appraisal theory Scherer and his colleagues proposed is called the "component process model," and as the name suggests, highlights different components of an emotion state, their coordination and integration, and the fact that they are not static states at all but unfold as processes in a particular sequence through time. Although there has been a historical assumption that appraisal requires cognitive, cortical types of processing, the theory is ultimately a functional model consistent with an evolutionary perspective that can certainly be applied to animals as well as to people and that can encompass cortical as well as subcortical structures.

A key feature of the component process model is that it makes testable predictions about the kinds of emotional responses (behavioral, psychophysiological, or, in humans, verbal) that result from each sequential stage of the appraisal process. For example, experiments predicated on the component process model pay close attention to how facial expressions change through time (within milliseconds), and how tone of voice can change quickly in time. In a sense, the emotion state is unpacked into a sequence of emotion states, each generated by a specific decision process that incorporates a new type of information. This emphasis on dynamics and time makes fMRI less good a method for testing the component process model. Instead, techniques with better temporal resolution, such as event-related potentials or magnetoencephalography (ERP, MEG; figure 4.2), offer a better fit, as do to some degree psychophysiological measures, video recordings of facial expressions, and spectro-temporal analyses of the voice.

Constructed Emotion Theory

Another influential and more recent theory of emotion, focused on emotional feelings and individual differences, has both psychological and neurobiological components. This is Lisa Feldman Barrett's theory of constructed emotion (also called the conceptual act theory) (Feldman Barrett 2017a). The theory subscribes to the idea of a core affect, characterized by dimensions of arousal and valence, and views these as low dimensional representations of interoceptive sensations. What is then added to this core feeling—the "constructed" aspect of the theory—are many of the contents we discussed earlier (concomitant sensory experiences, awareness of the eliciting situation, predictions about what might happen next—in short, everything of which you are conscious at that point in time). So any given conscious experience of an emotion includes core affect, bodily feelings, knowledge and thoughts about the situation that caused the feelings, and other conceptual representations. An important aspect of the theory is that memory and expectations are used in large part to construct the emotion—it is not simply caused by a stimulus. In this respect, the theory is in line with a large topic in neuroscience (so far mostly focused on sensory processing) that considers perceptual representations as arising from expectations and predictions. If one considers interoceptive sensations analogously, one view is that expectations and predictions about one's own body state (or the prediction error between this and the actual body state) constitute an important part of the conscious experience of emotions. This view has been elaborated by the cognitive neuroscientist Anil Seth (Seth 2013) and is also a component of Feldman Barrett's theory. The theory of constructed emotion is probably the most comprehensive attempt to incorporate all the rich content into the conscious experience of emotions that we often feel we have. Needless to say, it is strongly—perhaps exclusively—focused on human feelings, and so quite different from the functional approach to emotion states that we have taken in this book.

The theory of constructed emotion shares with appraisal theory an acknowledgment that emotions consist of many components that co-occur. However, unlike the component process theory, the theory of constructed emotion does not clearly separate feelings and concepts from other emotion components, and does not easily break processing down into distinct stages. Instead, it has a stronger holistic flavor, arguing that the very concept of the emotion is constructed in the brain by all the different processes that occur during an emotion—indeed, the theory appears to argue that the emotion state is just identical with the emotion experience, and furthermore that these are identical with the emotion concept, under an expanded definition of what a "concept" is. Learning, expectation, individual differences, and cultural effects are all strong points of emphasis in this theory.

One critical respect in which the theory of constructed emotion seems to differ from the other theories we are discussing in this chapter is that it argues that categories like fear, anger, and sadness are not "natural kinds" (categories we can discover in nature), but socially constructed categories that will vary by epoch and culture. It does not believe in "basic emotions," or indeed in brain systems of any kind for particular emotions, such as fear and disgust. Instead, according to the theory of constructed emotion, all emotion categories are constructed either in the minds of laypeople who rely on their cultural knowledge or in the minds of the psychologists and neuroscientists who are giving their interpretation to brain and behavior.

The theory of constructed emotion does not deny that there is some pattern in the brain that causes each specific emotion state. But it argues that the typology of emotions—the particular categories of emotions, like fear, disgust, and sadness—are constructed out of broader building blocks in the brain together with idiosyncratic contextual effects. Consequently, there are no systems for fear or anger, or indeed any categories, that one could search for in the brain. There is a strong emphasis on domain-general brain networks in this theory, which argues that the brain is not in any way specialized for processing emotions, but instead that there are many domain-general processes (like memory, perception, attention) that all work together to construct an emotion episode on the fly, as a particular situation requires it.

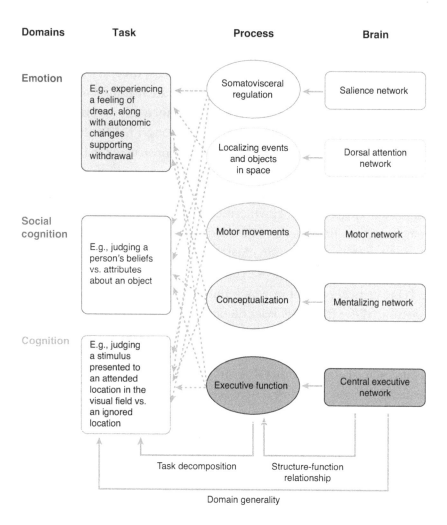

FIGURE 10.2. The theory of constructed emotion. In this scheme, the particular domain (e.g., "emotion") is related to particular tasks that can be used to infer particular processes, which are in turn implemented by particular brain networks. A notable feature of this scheme is the multiple mappings: there is no unique mapping from emotion to a single process or a single brain network, but instead emotion is "constructed" from many processes that are subserved by many networks and shared by other domains. Reproduced with permission from Feldman Barrett and Satpute 2013.

The theory of constructed emotion is one of the most ambitious emotion theories, as it tries to provide a comprehensive account of all aspects of emotion, from emotional behaviors, to concepts, to experiences (although, as noted, the emphasis is on the experience of emotion, and these different aspects of emotion are not clearly distinguished but rather seem to actually be identified with one another). It is also ambitious in that it incorporates a very rich corpus of modern neuroscience work, from detailed circuit-level data to large-scale networks as discovered from fMRI data. A flavor of the theory is shown in figure 10.2.

Constructed emotion theory is difficult to evaluate, and difficult to compare to what we have written in this book, since it is so broad and since it uses terms in different ways. Nonetheless, there are some clear points of agreement, and some clear points of disagreement (Adolphs 2017; Feldman Barrett 2017b). Many of the neurobiological themes that the theory stresses fit with what we have written here: unlike reflexes, emotions can involve learning and context. More than we have stressed it here, constructed emotion theory stresses the brain's ability to make predictions and to have expectations about the world as a key part of eliciting emotions.

We certainly disagree about one thing: that emotion states cannot have an objective, scientific taxonomy of emotion kinds. Constructed emotion theory denies that we can discover objective categories of emotions, like fear, anger, and so forth—it claims that there is no fact of the matter about what emotion categories are, and that they are just made up by people. As we noted, it will surely be the case that our current emotion categories will be revised, and likely will need to be subdivided. But we argue strongly that this is an empirical task of scientific discovery, not a process of social construction where we can just make up any emotion categories we like. In this respect, the multidisciplinary scientific discovery of emotion categories is no different from the discovery of other scientific categories: like galaxies, planets, atoms, or molecules, these are ways of carving up the world based on the best scientific measurements and theories we have available. They are certainly always works in progress, but they aim toward an objective description of the world that goes beyond the concepts and categories we use in everyday life.

One way of highlighting the difference between constructed emotion theory and our view is to imagine scientists in different cultures, or even alien scientists, investigating emotions. Constructed emotion theory assumes that these would all produce their own, different emotion categories, a view supported by historical and cross-cultural observations.

We assume that they would end up producing the same emotion categories, because we believe that science has to approach a consensual representation of the world (even if there are differences in detail and terminology). Just like alien scientists would also have theories that include galaxies and planets and atoms, they would also have theories that include emotions, and emotion categories and dimensions that a future science of emotion on earth will produce.

Jaak Panksepp's Emotion Systems

A strong stance against most of the views of the theory of constructed emotion comes from the late Jaak Panksepp's neurobiological systems theory (Panksepp 1998), even though his writings predate the theory of constructed emotion. Based largely on older animal studies involving methods like electrical brain stimulation (much of it Panksepp's own seminal work), this theory proposes a handful of core emotion systems that operate in humans and animals. It is thus like Paul Ekman's view of basic emotions—it proposes that there are specific modules corresponding to specific categories of emotions. It stresses the important role of subcortical structures, and it is entirely unabashed in its belief that all animals that have emotion states also have conscious experiences of the emotions. For Panksepp, the neurobiological and behavioral data are entirely sufficient to attribute subjective feelings of emotions to animals; verbal report is seen as mostly irrelevant in this regard.

Whereas Ekman had proposed six to seven basic emotions from human facial expressions (happiness, surprise, fear, anger, disgust, sadness, contempt), Panksepp proposes seven basic emotional systems rooted in his observations of animal behavior and neuroscience: seeking, rage, fear, lust, care, panic, and play (in his writings, these systems are denoted by all capital letters). A large part of the empirical evidence for subcortical neural substrates for these systems, according to Panksepp, comes from electrical stimulation and neurochemical studies. His view is also substantially inspired by homologies between brain structures across mammals (and to some extent all vertebrates), homologies that are more prominent for subcortical structures than for cortical regions.

The strongly modular view of Panksepp's theory is attractive in that it poses neural circuits that can process many different situations to coordinate the same emotional reaction. It provides us with a "vocabulary" of emotional repertoires that can be engaged across different situations. This idea is similar to our emotion property of generalizability we listed in figure 3.2, and the image of emotion states as showing a fan-in, fan-out architecture that can link the same emotion state to very many different stimuli through learning or evolution. Once we have a central module for coordinating fear responses, this can be applied to a nearly unlimited range of stimuli that acquire threat meaning through experiential learning, observation, or instruction. Nonetheless, it is likely to be the case (from the neurobiological data we reviewed in earlier chapters) that "fear" will turn out to be processed by several functionally related, but neuroanatomically distinct systems, and that each of these systems will have a more restricted range of stimuli about which it can learn. So there will probably be more than just seven emotions, or at least each of the seven will have several subtypes.

It is important to emphasize that, according to Panksepp's view, you can learn what situation to be afraid of, but you can't learn a totally new emotion state. The emotion states are set by evolution and encapsulate the recurring environmental challenges and themes that animals have encountered over generations. The flexibility of emotion states allows these states to be linked to many stimuli and behaviors through learning, but the states themselves are species-specific and new ones cannot be learned.

Like Damasio, Panksepp emphasizes subcortical structures as playing a critical role in emotions. Whereas Damasio's theory tends to emphasize interoceptive representations in the discussion of feelings (that is, sensory representations), Panksepp's view tends to emphasize motivations and behavior (that is, motor-related representations). Both Panksepp and Damasio tend to emphasize subcortical regions more than does Feldman Barrett, and both stress cross-species data and acknowledge homologies across different species.

We find Panksepp's view attractive, although we have doubts about the specifics. The emphasis of the theory on careful ethological observation and the cross-species emphasis that prominently includes animal studies are attractive features (indeed, most of the data in Panksepp's book comes from experiments in animals, whereas most of the data in Feldman Barrett's book comes from experiments in humans—needless to say, each with very different kinds of dependent measures). We have some doubts about the specific list of emotion categories and their neural systems—but this is not really saying much more than acknowledging that we just need a lot more neuroscience studies, using modern and future methods, to uncover these. We also share Panksepp's belief that animals have conscious experiences of emotions, but unlike Panksepp, we don't feel confident that we have a good way to investigate this yet.

Edmund Rolls's Theory of Emotion

Another neurobiological theory of emotion that is worth noting comes from Edmund Rolls, whose work at Oxford University, mostly in monkeys, provided many of the foundational studies for how reward and punishment are represented by single neurons in the brain. Rolls and his lab had found neurons in monkeys, as well as brain regions in human fMRI studies, that responded to the value of stimuli independently of their sensory properties. For instance, the orbitofrontal cortex and the amygdala responded to chocolate in somebody hungry for chocolate, but stopped responding to chocolate when the person was satiated and no longer wanted to eat chocolate. This finding has now been replicated many times, using fMRI and using single-neuron recordings, and is considered a hallmark of value representations that are sufficiently flexible to support goal-directed instrumental behaviors. Like emotions, these flexible value representations are "decoupled" from the stimuli. They are encoding something more abstract—what the stimuli mean, what value they have—which changes depending on the homeostatic state of the animal. Also like emotions, such representations motivate adaptive behavior—wanting to obtain and eat chocolate or not.

Rolls proposed a theory in which the emotion states are tightly linked to behavioral motivation, whereas the emotion experience is more dependent on language. Rolls's theory featured an interesting two-dimensional space in which emotion states are situated, which maps emotion categories in terms of the behaviors elicited by the administration or withholding of rewards and punishments. For instance, the administration of reward would map onto happiness, its withholding onto frustration; the administration of punishment would map onto fear, its withholding onto relief. While it is difficult to see how this space alone could provide enough distinctions to capture the diversity of emotions, it is an interesting behavioral alternative to the usual psychological dimensional spaces (which, as we have noted, typically take arousal and valence as the two fundamental dimensions).

Joseph LeDoux's Higher-Order Theory of Emotion

We already mentioned LeDoux's view that emotions should be considered as conscious experiences, and that the animal studies of emotion states we reviewed in chapter 6 are about what he terms "survival circuits," which are not emotions according to his view. In his words,

> The biological systems that underlie these varied states are not there to make fear. They exist to keep the organism alive. When the organism has a brain that can be conscious of its own activities, activation of a biological survival system contributes to fear. We can study biological survival systems in animals. But fear—the conscious experience—can only be studied in organisms that have the cognitive wherewithal to be conscious of these activities. (LeDoux 2017, personal communication)

It is important to note that LeDoux does not deny that animals might be conscious. But the kind of consciousness of which they are capable is more limited and does not extend to self-awareness. Animals might have conscious feelings when they are in a state of fear, but they do not know that they are afraid. By contrast, humans have an extended sense of self, and when they are afraid, they know they are afraid—and consequently they can tell us about it. More importantly, LeDoux argues that the conscious experiences of emotions are not "products" of subcortical circuits—structures such as the amygdala and the hypothalamus, which we reviewed in chapter 6.

A part of our disagreement with LeDoux is just semantic. Whereas we define emotions as functional states, and then consider the question of their conscious experience as something further to be empirically investigated, LeDoux instead defines emotions as conscious experiences from the start: "we define 'fear' as the conscious feeling one has when in danger" (LeDoux and Brown 2017). Most psychological work on emotion probably uses the same working definition—but conflates it with behavioral, neural, or even functional approaches without further clarification. In a sense, LeDoux's definition is just a helpfully explicit statement of what many other views assume implicitly anyway (this is precisely the problem that LeDoux diagnoses [LeDoux 2015; 2017]).

LeDoux's higher-order theory of emotion experience (which is more a theory of consciousness in general than it is a theory of emotion) is economical and provides specific components, which make neuroanatomically testable hypotheses. The basic idea is that there is a single, general cortical system for producing conscious experiences. This system has layers that arise from more or less elaborate representation of a self, an idea that Antonio Damasio first proposed. It also requires attention, metacognition, and working memory, so that the representations that give content to a conscious experience can be "held in mind" and thought about. In this view, the same cortical system for producing conscious experiences just gets different inputs that determine what it is that we are conscious of on any occasion. If they are visual inputs, we are conscious of seeing something. If they are auditory inputs, we are conscious of hearing something. If they are inputs from subcortical "survival circuits," we are conscious of feeling an emotion.

Aspects of LeDoux's scheme actually fit fairly well with ours, differing mostly in terminology and emphasis. Like LeDoux, we do not believe that emotion states confined to subcortical systems are somehow intrinsically conscious. We also share the belief that interaction with a number of other cognitive processes (attention, working memory) will be required in order to have a conscious emotional experience. Where we differ is in the extent of processing that is allowed for an emotion state. If one restricts the state only to subcortical processing, and only to processing that produces behavior, then an emotion state may have little immediate connection with a conscious experience. If one extends the emotion state to include also cortical processing and interaction with other cognitive processes (as we do), then a conscious experience of the emotion may follow as a causal consequence of (or even as constituted by) the emotion state. However, we emphatically believe that subcortical structures are necessary for the conscious experiences of emotion, even though they may not be sufficient. Most importantly, we believe that emotions have an evolved biological function and are not mere epiphenomena of higher-order neural processing that can keep cognitive neuroscientists busy.

Philosophy of Emotion

Given the heterogeneity in the theories surveyed above, given the deep conceptual issues, and given the close link to consciousness, it may come as a surprise that philosophers haven't contributed more to theories of emotion. One might think that the field could really benefit from some clear philosophical exegesis. But somehow, emotion has eluded most of analytical philosophy, and what philosophy of emotion there is has generally tried to analyze our concepts for feelings and emotions, rather than to connect strongly with cognitive psychology, let alone neuroscience. There are some exceptions. The philosopher Jesse Prinz has led interesting arguments, somewhat in the vein of James and Damasio, about the role of bodily states in representing emotional themes (Prinz 2004). Somewhat more tangentially to emotion per se, the philosopher Alvin Goldman has argued that we figure out how other people feel by simulating their bodily states (Goldman and Sripada 2005). These theories draw on neuroscience data, such as the finding that experiencing an emotion, and attributing it to another person from their facial expression, seem to involve overlapping brain regions (chapters 8 and 9).

The philosopher Andrea Scarantino has surveyed theories of emotion as falling largely into three approaches that take different aspects of an emotion as foundational or primary in some sense. These are appraisal, feeling, and motivational approaches, which, respectively, emphasize the evaluative, experiential, and behavioral components of an emotion (Scarantino 2016). Scarantino's own approach actually shares some components with ours. His view, called the motivational theory of emotion, derives partly from work by the psychologist Nico Fridja and sees emotions as tendencies or dispositions for action (Scarantino 2014). Like the list of features we provided, Scarantino locates emotions as motivating behaviors while exhibiting certain phenomena: flexibility, impulsivity, and bodily underpinnings.

One of the largest impacts of philosophy on the science of emotion has come from Paul Griffiths's seminal book, *What Emotions Really Are* (Griffiths 1997). Griffiths argues persuasively that different emotions fall into several broad types, that distinctions need to be made between the phenomena commonly lumped under affective science, and that no single theory can do the job. Notably, Griffiths argues that not all the words and concepts we have for different emotions will be scientifically useful; that is, that they will not all correspond to natural kinds that the scientist can discover.

We think that much of this is right, although we also think that much of it is an empirical question. It may indeed be the case that the rodent neurobiologist conducting optogenetic studies of aggressive behavior, and the social psychologist interested in people's feeling of anger, are just studying very different phenomena. That may require a revised taxonomy of emotions, a dimensional view, or simply an acknowledgment that in the human case there's a lot more going on in addition to what's going on in the rodent case. None of these three possibilities should seem particularly surprising. After all, even if the circumstances are made as similar as possible, the way that a human can cope with a situation that specifies anger will obviously be different in many ways from the way that the rodent can cope. So the functional relationships that define a case of rodent anger will almost always look very impoverished compared to the human case. That does not necessarily mean that "anger" (or RAGE, in Panksepp's terminology) is not a natural kind, however. It may just mean that "anger" has a simpler, but related, function in a rodent compared to its elaborated role in humans. Without a lot more empirical detail to the story, we do not know whether we should make a completely separate category for "human anger" as compared to "rodent anger," or whether we should consider them subtypes of a single larger functional category because of similarities that they share. Neuroscience data would seem to offer a particular compelling source of data here, since there are clear brain homologies between the two species. Even if some of the behaviors are necessarily rather different, since the ecology of these two species is different, the brain systems for anger may look similar.

One important emphasis in Griffiths's work has been to view emotions, at least in humans, as strongly social. Indeed, even Ekman's "basic emotions" are explicitly social—they are derived from social communication through facial expression. So are most of Charles Darwin's examples—they are behaviors of mammals that we can observe and interpret as emotional. Darwin thought of these behaviors as reflecting "serviceable (useful) associated habits," and believed one could understand the true emotion behind the behavioral display by going back to how it had evolved. But clearly many emotional behaviors, especially in humans, also subserve explicit social communicative functions (such as the different functional roles of a smile we mentioned earlier in the book; Rychlowska et al. 2017). For that matter, the social function of many types of emotion states may not be derivative to any kind of "serviceable associate habit" that precedes it; emotions like pride, guilt, shame, and embarrassment plausibly evolved only in the service of a social function.

This is an important point that greatly complicates the study of emotion, since we now need to interpret emotional behaviors not only "in the light of evolution" with respect to the function that they ancestrally played (for example, wide eyes and bared teeth for fear subserve functions in maximizing the visual detection of threats and prepare the animal to defend itself), but also with respect to the function that they play in the regulation of social interactions (for example, the same fear display also serves to signal fear to a predator or a conspecific) (see box 2.3). Once emotional behaviors were co-opted for social communication, this of course bootstrapped further adaptations: those that explicitly use emotional signals as social signals or warning signs, and those that might even use such signals deceptively. The role played by emotions in social communication is a huge topic, and we would need to write a whole other book to do it justice.

Griffiths argues that emotions are a heterogeneous lot, and he seems to think this is so in good part because the functional relationships into which they enter are to some extent flexible. Remember way back in chapter 2 we gave the example of clocks as functionally defined devices. But, we noted, a clock could also be used as a paperweight, or perhaps the ticktock of a grandfather clock could be used as a sleep-inducing device for cats. Those examples were intended to be in jest—they were intended to illustrate the point that clocks are designed to keep time, and so other uses of them are in a sense mistakes or malfunctions and do not reflect their proper function. In the case of emotions, the idea we proposed was that evolution had selected particular functions as the "proper functions" for emotion states. This is a widely held view about biological functions in general, argued in the most detail by the philosopher Ruth Millikan (Millikan 1989). The heart's function is to pump blood; using a heart as a teaching device in medical school does not make the heart a teaching device because that is not its proper function. The lung's function is to exchange oxygen and carbon dioxide; using it to blow out a candle does not make it an air blowing device because this is not its proper function.

These last two examples may strike you as not quite equivalent. Yes, the heart is certainly not a teaching device. But maybe the lungs are a blowing device of sorts? We seem to use them that way quite a bit. What if emotion states are more like lungs than like hearts in this respect? What if emotion states don't have a single proper function at all, because they have been used throughout evolution for whatever purpose is at hand for which they were most useful? Sometimes a state of fear is useful if you want to avoid being eaten by a predator. Sometimes it is useful just to signal submission to a dominant member of your own species. Sometimes it's even useful if you want to entertain people by being a method actor. What if, that is, emotions, perhaps especially in humans, are so flexible that they can be applied in a very large number of different contexts? That would make it hard to discover what they are really about, and it would make it look as though they are not "natural kinds" at all. This may also be the view that underlies Feldman Barrett's constructed emotion theory.

Our response to this possibility is that we would need some actual data, some studies that could inform these questions. Possibly, developmental studies, comparative studies, and careful ecological work would help. However, we also think that neuroscience would help. In particular, if the neuroscience data showed us that all the different uses to which a state of fear is put share a common substrate, we could have a mechanistic explanation that shows how a core emotion state can subserve multiple functional roles. This might lead us to shift our foundations for thinking about emotion from functional criteria (as we have done in this book) to mechanisms. Theories like Jaak Panksepp's seem to bet that we would find such core emotion systems; theories like Lisa Feldman Barrett's seem to bet that we would not. We think we need to do the neuroscience to find out.

Taking Stock

In humans, feelings and emotion states are so closely intertwined, and feelings are so salient in our lives, that it is little surprise that most theories of emotion focus on feelings. One might well ask if there are any clear effects of emotions in the absence of feelings. For sure, there are no cases of somebody not having a conscious experience of an emotion yet showing a strong, comprehensive emotional reaction (for example, screaming and running away from a bear without feeling fear); on the other hand, it would make little sense to presume this is even possible, since consciousness is required for all the components. Piotr Winkielman and Kent Berridge have argued that there are unconscious emotions, and there are well-documented cases where emotional stimuli have effects on attention or decision-making even when they are not consciously perceived—at least for relatively faint emotions (box 10.1).

The many theories of emotion, which we only skimmed here, can themselves be arranged along some dimensions. Some, like Damasio's, are strongly foundational in trying to provide a fundamental basis for what emotions are all about in the first place, whereas others, like Feldman Barrett's, are more descriptive and assume certain phenomena as given (such as the experience of emotions). Some attempt to reduce emotions to their ingredients, such as Rolls's view that emotions are fundamentally states linked to reinforcers, whereas others (especially many psychological theories) tend to be more holistic.

There is no shortage of theories of emotion, and there is no shortage of emerging neuroscience data on emotions. What is missing are
data that could actually help us decide if a theory is right or wrong (or
needs modifications). This, in turn requires work on the part of the
people who come up with the theories and on the part of the people
doing the empirical research. The theories typically need to be more
circumscribed in their domain and more fleshed out in their details, so
that they would make specific testable predictions. It needs to be clear
what kinds of experimental results would provide support for a theory,
and what kinds of experimental results would falsify a theory. Similarly,
the experiments need to be carefully planned so that the variables that
are manipulated and measured have construct validity: they need to
correspond to terms in the theory, they need to be measured reliably,
and the logic of the experiment needs to provide data that support its
conclusions. This is a tall order and requires close discussions between
the people formulating the theories and the people doing the research.
In the next and last chapter, after we have summarized the key points
of this book, we give a few examples of future projects that could help
bring theories and experiments closer together.

BOX 10.1. Are there nonconscious emotions?

There is a long literature on the possibility that we can be unaware
of aspects of our own emotion. Going back to Freud's concept of
the unconscious mind, there is the idea that we might repress, or
otherwise lose access to, how we feel. We may also be unaware
of the reasons for why we feel an emotion. It is important to distinguish two cases:

1. *Unawareness of stimuli that induce an emotion state.* There is
    good evidence that sensory stimuli can go undetected yet still
    induce an emotion. For instance, faint odors may induce a
    subtle sense of disgust without us being aware of the odor. A
    more stringent test asks subjects explicitly whether they can
    detect the stimuli. Here, too, clear effects have been found.
    For instance, visual stimuli can be presented so that they

are processed to some extent, but not consciously detected (through techniques such as backward masking or continuous flash suppression, for instance). When highly emotional stimuli are shown in this manner, they can still have some effects. Subliminally presented pictures of nudes can capture visual attention, and this effect depends on the gender orientation of the perceiver: heterosexual men show the strongest effect for pictures of nude females, gay men show a weaker effect, and heterosexual women a weaker effect still (Jiang et al. 2006). Another effect is on decision-making: a valence-specific effect on the ability to detect the direction of moving dots could be produced with subliminally presented emotional pictures (Lufityanto, Donkin, and Pearson 2016). In this latter study, the experimenters used a simple linear accumulator model (conceptually similar to the drift-diffusion models we discussed in chapter 3, but simpler and without noise), and found that the ability to detect the direction of moving dots (a common psychophysical task) could be improved with valence-specific emotional images (a snake) that were shown masked so that they could not be consciously perceived.

There are also nonconscious influences of higher-order stimulus properties on emotion. In Pavlovian fear conditioning in humans, for example, it seems that conscious awareness of the temporal relationship between the US and CS is not required for conditioned autonomic responses to emerge, a form of emotional learning. However, in all these examples there is not an intense emotion state involved. To our knowledge there are no cases where a strong, full-blown emotion is evoked by a stimulus, and that stimulus cannot be consciously perceived. Nonetheless, there are of course pathological cases where a strong mood, such as depression, can be induced without the patient knowing why they became depressed; analogous effects can also be produced by direct brain stimulation or drugs.

2. *Unawareness of the feeling of the emotion.* Normally, emotion states cause or are otherwise accompanied by feelings. In a

clinical condition called alexithymia, patients appear unable to describe how they feel. It is somewhat unclear whether they really do not experience any feelings, or whether they have difficulty expressing their feelings in language. At the opposite end of the spectrum is increased "emotional granularity," the ability to report on many finely differentiated feelings, which is correlated with resilience to depression and anxiety (Kashdan, Feldman Barrett, and McKnight 2015). There are also claims that humans can have emotional feelings that are just plain unconscious—valenced states that are evident in behavior, but not in verbal report. Piotr Winkielman and Kent Berridge did an experiment in which subjects were shown subliminal pictures of emotional facial expressions (Winkielman and Berridge 2004). They were unaware of the faces that they saw, and seeing the subliminal faces did not change their self-reported affect. Yet there was an effect on how much they chose to drink of a pleasant beverage, arguably a nonconscious emotion effect on this behavior. Interestingly, similar dissociations have been reported following brain damage in a patient: that patient was unable to rate the pleasantness of drinks (salt water, lemon juice, and sugar water all yielded the same pleasantness judgment, and even produced the same pleased facial expressions when consumed), but he strongly preferred the sugar water over the other two when he was given a behavioral choice (Adolphs et al. 2005b). Once again, however, these studies dissociate rather narrow aspects of an emotion; there is no case of a strong, comprehensive emotion state induced in a healthy human under conditions where they report feeling no emotion.

Summary

- Most emotion theories focus on the conscious experience of emotion, which encompasses several types of conscious content.

- Most psychological theories of emotion propose that conscious experiences of emotion have "core affect"—an underlying two-dimensional structure of valence and arousal.

- Other emotion theories include those of Damasio, Panksepp, Scherer, Rolls, Feldman Barrett, and LeDoux. All of them prominently discuss the conscious experience of emotion.

- Some philosophical work on emotion suggests that emotions may have very flexible functional roles, posing a challenge to how to study them as natural kinds.

CHAPTER 11

Summary and Future Directions

This book aims to provide a foundation and a framework from which testable hypotheses can be motivated and investigated with modern neuroscience tools in people and animals. It is based on a few strong but widely shared premises, makes specific claims that are based on the evidence so far, and proposes strategies for the future. If you take away nothing else from this book, here are the major points that underlie our view of how to understand and study emotions.

The Main Points of This Book

1. Emotions are fundamentally biological phenomena; therefore, they need to be understood in biological terms. Our proposal is that answering what emotions are about requires descriptions of their biological function.
2. Emotions are ubiquitous across species and evolved by natural selection; therefore, they should be studied and compared across animals and humans. Charles Darwin already emphasized points 1 and 2.
3. We need to keep distinct three broad aspects of emotion that are often conflated: the emotion state, the conscious experience of the emotion, and thinking about the emotion (using concepts and words). This book is mostly about the first of these.
4. Emotions can be studied without needing to study subjective feelings. A science of emotion can therefore proceed without needing to measure verbal reports or needing to tackle theories of consciousness.
5. Emotions are implemented by neural mechanisms that we can discover and manipulate with neuroscience methods. Studies in

mice are revealing that even relatively small brain regions, such as the central nucleus of the amygdala, do not exert a unitary function but rather contain multiple neuronal cell types that can exert different influences (even in opposite directions) on a given emotion state. New tools such as optogenetics and calcium imaging are therefore extremely important for a neuroscience of emotion.

6. Emotions have characteristic properties—building blocks (or "emotion primitives") and features—that provide objective criteria for studying them (figure 3.2). Emotion properties allow us to identify instances of emotions across species and could provide a dimensional space in which we can compare and categorize different emotions.

Perhaps the strongest single premise in our book is the first point: that emotions are fundamentally biological phenomena. We currently know of emotions, however you conceive of them, only in biological organisms. Hurricanes, protons, stars, and cars do not have emotions. This means we need to understand and investigate emotions as biological phenomena: how did they evolve, what adaptive functions do they subserve, and what are the neurobiological mechanisms that achieve this?

The view that emotions reflect evolved adaptations is widely accepted among both biologists and psychologists, and features prominently in Charles Darwin's book, *The Expression of the Emotions in Man and Animals* (Darwin 1872/1965). Darwin's argument that emotional expressions are evolutionarily conserved among phylogenetically related species is based on two fundamental observations. First, expressions of specific emotion states, such as those that motivate defensive behavior (for example, "fear" or "anxiety"), exhibit similar features in related species. Second, the nature of these expressions, that is, the particular form that they take, can be rationalized in terms of the original function—the utility—that they provided for that species relative to its environment. Thus, for example, when threatened by a predator or aggressive conspecific, many animals bare their teeth. This served an original function to display the weaponry the animal would bring to a physical conflict, and thereby deter such a conflict. However, not all emotional expressions have an obvious functional utility for the context in which they are exhibited. For example, Darwin cites the example of the apparently useless behavior exhibited by a cat that kneads a blanket with its forepaws, a behavior that originally served to stimulate the flow of milk from the mother when the animal was nursing as a kitten. Because the behavior is tightly linked to an underlying emotion state, Darwin argues, any stimuli that trigger that emotion state will elicit the kneading behavior—whether it has any utility or not. These and other instances are described in his first chapter on the "Principle of Service-able Associated Habits." This type of phenomenon does not allow us to infer *which* emotion state is being elicited, but it is strong behavioral evidence of an associated emotion state of some kind.

Our evolutionary view of the functions that emotions subserve in turn led us naturally to include animals in a science of emotion, just like it did for Darwin. Even if human emotions did not have a clear evolutionary ancestry, studying the varieties of emotions in other species would be informative, because this would give us a broad view on how nervous systems have solved fundamental ecological problems. There are also strong practical reasons for taking a comparative approach to the study of emotion. There are experiments that are impractical or impossible to do in humans, and techniques with an essential level of causal manipulation or anatomical resolution that can only be applied in animals. A science of emotion would be hugely handicapped, and hugely biased, if it restricted itself to human beings. Compare our knowledge of vision, or of memory: these fields have benefited tremendously from a multimethod, multispecies approach. We can relate visual studies using fMRI in humans to circuits and single-unit recordings in monkeys and rodents and flies. We need the same rich approach for a science of emotion.

Emotion states are distinct from experiences of emotions, concepts of emotions, and words for emotions. An emphasis on experiences or concepts of emotions in cognitive neuroscience reflects the anthropocentric bias of many emotion studies. But these are not directly studies of the emotion itself, and they are not necessary to study the emotion. We can study emotion states without needing to study emotion concepts or emotion words. Indeed, verbal reports of conscious experiences, what people say they think and believe about emotions, and the words that laypeople use to talk about emotions, are objects of study in themselves. They can inform us about cultural variability, about the development of emotion concepts, and about differences between languages. But they should not be conflated with the study of the emotion states, which need to be discovered by biological methods applied to animals as well as humans.

Importantly, we are not advocating that the study of emotions in animals should be restricted to, or even focused on, determining whether a given species has an internal state that is equivalent to a particular human emotion (for example, "fear"). Rather, it should be aimed at applying objective criteria to determine whether a particular set of phenotypic expressions, measured in an animal under a given set of conditions, displays the core properties of an emotion state (for example, those we listed in figure 3.2). If it does, it is then an interesting further question whether that state is homologous or analogous to similar states that we can find in humans using similar dependent measures. If that state is triggered by exposure to a predator and involves behaviors like hiding, escaping, and defense, it is a semantic issue whether we call it "fear" or a "defensive emotion state." What matters is to understand the brain mechanisms that control the different properties of this state— such as its valence, its scalability, its persistence and its generalizability.

Feelings Again

While we omit feelings (conscious experiences of emotion) from most of the book (but see chapter 10), and while we argue that this aspect of emotion is inessential to study emotion states scientifically, we nevertheless personally believe that animals have feelings. This is also the conviction of many laypeople and other scientists, and there is no shortage of books on animal feelings. Marc Bekoff has written a book with the title *The Emotional Lives of Animals* that is all about animal feelings (Bekoff 2007). Carl Safina's book, *Beyond Words: What Animals Think and Feel*, similarly provides a wealth of anecdotes that persuade many people to attribute feelings to animals (Safina 2016). There is even a new journal featuring many articles on animal feelings, called *Animal Sentience* (http://animalstudiesrepository.org/animsent/).

Unfortunately, there is no consensus on whether behavioral evidence is sufficient to conclude that animals have conscious experiences of emotions. Until we have such evidence, a science of emotion in animals that includes conscious experiences of emotion remains problematic, because there will be vigorous disagreement on this issue. Even once we have that evidence, a science of emotion in humans and animals can be productive without having to tackle difficult problems about conscious experience. It is for these reasons, as we discussed throughout the book, that we believe that a science of emotion should for now separate the study of the conscious experience of emotion from the study of the emotion state. Once again, we can look to other successful scientific examples that have done the same: vision science, for instance, does not require the scientist to be studying qualia (for example, the subjective experience of seeing red versus blue)—that's a different, and much more difficult problem, than understanding vision. We should do the same thing with a science of emotion: separate emotion and the conscious experience of emotion. Once this starting point is chosen, it is unproblematic to include animals in a science of emotion (and for that matter those human beings for whom we also lack verbal reports, such as infants, demented patients, or patients who are aphasic).

It is worth noting the inconsistency of a standard argument one often hears in favor of the idea that humans but not animals have conscious emotional experiences. The question is this: if you doubt that animals can consciously feel emotions, then why don't you also doubt that other people can consciously feel emotions? The standard argument one hears is that other people are like oneself, and so one can make this inference validly by analogy, whereas animals are too different. But take the case of a preverbal child or a person who speaks a language you do not understand. You can tell a lot about how they feel from their situation, their body posture, their behavior, even without verbal report. Well, that is exactly the same kind of evidence you have for the animal. Importantly, this is not the same evidence as you use to assess how you feel, in your own case; you decide how you feel based on introspection, not by looking in a mirror. Therefore, the argument by analogy doesn't work.

In your own case, you know you feel afraid without needing to observe your own behavior. In another person's case, you can only know how they feel by observing their behavior (or listening to what they report that they feel). Those are very different types of data, so what makes you think that the other person is feeling what you are feeling (or indeed that the behaviors you see in another person are related to feelings at all)? We don't want to digress here, but we raise this point to suggest not only that it is perfectly valid to study emotion states in animals, but that it may even be possible in the future to study the conscious experiences of those states in animals. There is no good justification for the asymmetrical belief that conscious experiences of emotions occur in other people, but not in animals. Put another way, claiming that one can only study the "subjective experience" of emotions in humans, and then doing that by studying subjects other than oneself is no different from claiming one can study the subjective experience of emotion in animals.

Future Experiments

A primary reason why we have devoted considerable space to describing overarching concepts and themes in this book is very simple: it is still very early days for a science of emotion. Whatever data we describe right now will likely need some revision by the time this book is published, and might well be premature in trying to support strong conclusions. We hope we conveyed some sense of how much we still have to learn about the neural circuits that instantiate emotion states. On the other hand, the techniques, and the analytic logic of modern neuroscience studies on emotion, offer some very powerful tools that we can use to think about what we might do next.

We thus end this book with a brief list of open questions. Armed with more sophisticated concepts, clearer terminology, a survey of the modern methods available to neuroscientists, and a sense for the brain circuits and systems revealed in studies so far, we can try to formulate experiments that might try to answer some of those initial questions. Another reason we are ending with some future experiments is that they make the most specific recommendations for what might still be missing. In the experiments we imagine below, we identify some specific lacunae—techniques and approaches that are required to enable the experiment, but that are not yet quite available.

We begin by sketching some experiments that could nearly be done, then move on to more challenging "experiments" that are really just open future questions, challenges that a science of emotion may eventually confront.

Open Question 1. Does Activity in Subcortical Structures Like the Amygdala Produce Emotion States?

Hypothesis. Electrical or optogenetic stimulation of subcortical structures in an intact animal or human brain is sufficient to reproduce many or all measurable effects of an emotion.

Support for or against this strong hypothesis would go a long way toward resolving some current debates in the neuroscience of emotion. Surprising as it may seem, there is ongoing debate as to whether activity in subcortical structures causes emotions. On the one hand, as we reviewed in chapter 6, selective stimulation of specific neurons in subcortical structures such as the amygdala and the ventromedial hypothalamus is sufficient to instantiate many of the properties of a defensive emotion state, in rodents. But the evidence is much murkier in humans. Stimulation of subcortical regions in the human brain has indeed been reported to evoke verbal reports of the conscious experience of fear or anxiety in humans, for example. However, those results are largely anecdotal and have been reinterpreted by some as reflecting nonspecific "arousal," rather than a true "emotion" (LeDoux 2015), due to the difficulty of restricting electrical stimulation to just one brain region with the techniques used at the time.

So, how could we resolve this impasse? The ideal experiment would be to precisely stimulate those neuronal subpopulations in subcortical structures that we hypothesize to be sufficient for eliciting an emotion state, in both humans and in animals. It would be important to do this using technology that does not stimulate other neighboring neuronal populations, or fibers of passage (axons that are coursing nearby). That level of precision can be achieved using optogenetics in nonhuman animals. However, this experiment cannot yet be performed in humans, for both ethical and technical reasons (box 5.5). At present, the best one can do in humans would be to use focal brain microstimulation.

The effects of these stimulation trials, in both humans and animals, should not be evaluated by a single type of dependent measure, but should be evaluated using a battery of dependent measures that include parameters than can be observed in both humans and animals, such as behavioral observations, physiological and endocrine measures as well as verbal report (in human subjects), and, importantly, whole-brain measures such as fMRI (in both humans and animals). We would want to look at the activation of homologous brain structures across species, so in this experiment we would restrict ourselves to mammals. One way to craft such a study would be to ask a progressive series of questions like these:

- Is electrical stimulation of the amygdala (or any other subcortical brain region of interest) in humans sufficient to cause conscious experiences of emotions as measured by verbal report?
- If yes, then what is the accompanying global pattern of brain activation as measured with fMRI?
- Do we see a similar pattern of brain activation measured with fMRI in a monkey or a mouse during optogenetic stimulation of a specific neuronal subpopulation in the amygdala? Do humans have this same neuronal subpopulation in their amygdala?
- Do we see other behavioral and physiological effects of an evoked central emotion state in these cases?
- Does this stimulation induce a state that exhibits the basic properties of an emotion, including persistence, scalability, valence, and generalization?

In this way, we would try to obtain as much evidence for the recreation of a central emotion state as possible, through multiple dependent measures, and through observation of the properties of the emotions we listed in figure 3.2. The question is whether there are similar patterns of global brain activity that are evoked by specific stimulation of a subcortical structure or cell population in humans and animals, and, if so, whether these patterns are correlated with common dependent measures of an emotion state that can be evaluated in both species.

Limitations. There are at least three limitations to this experiment. First and foremost, there are clear methodological challenges—challenges that point the way to new techniques that we will need to develop. We would need to be able to have considerably more precise stimulation in humans than is currently possible, ideally equivalent to optogenetic stimulation. This is not impossible in principle, but the experiment would always be limited to rare human patients in whom there is clinical justification for invasive implantation of optic fibers or electrodes, as well as genetic modification of neurons using viral vectors (box 5.5).

A second limitation, also largely methodological, is that we would need to come up with a more mature and more complete list of what to look for as evidence of an emotion state. What counts as reproducing "all effects of emotion"? One way to address this concern might be to compare the experimentally evoked emotion to a naturally evoked emotion. If we confront a mouse with a predator and quantify as many dependent measures as possible, we could draw up a list and ask: of all the effects of the emotion we see under natural circumstances, which are reproduced with the electrical stimulation?

This second limitation immediately points us toward a third limitation, the fact that the electrical (or optogenetic) stimulation is artificial. We are clearly not mimicking what would happen in the brain under natural circumstances in the experiment just described. One step toward addressing this concern would be to optogenetically stimulate amygdala neurons in precisely the same pattern that they would normally show activation during a naturally evoked emotion state, essentially replaying the pattern of activity that we might have recorded in the paragraph above with a real predator. This has in fact been done, in studies of memory in mice, so it is possible in principle. However, even such a precise replay is still an artificial brain state. We could include other subcortical structures in our set of neurons to activate with precisely patterned stimulation. But no matter how precisely we replicate the spatiotemporal pattern of neuronal activation in a restricted set of subcortical structures, the fact remains that the animal is not actually smelling or seeing a predator. Nevertheless, there are anecdotal cases where stimulating a particular brain region in a human can evoke a sensation or percept as vivid as if it were triggered by an actual external object, so in principle this could occur in an animal as well. By stimulating subcortical structures, we can ask about the minimally sufficient subset of an emotion state—but we would need to keep in mind that we are leaving out some components.

Open Question 2. Do Subcortical Structures Cause Global Emotion States by Also Recruiting Other Necessary Structures?

Hypothesis. The induction of an emotion state through subcortical stimulation also requires other brain structures, including cortical structures in some species.

The question asked in this experiment builds logically on the Open Question 1. We assume that an emotion state requires processing in a distributed system of several interacting brain regions. As such, stimulation of only a subcortical node, such as the amygdala or the hypothalamus (or both), might be sufficient to induce an emotion state, but we would hypothesize that it would do so in good part by recruiting many other brain regions. Which of these other brain regions are also necessary for the emotion state?

To answer this question, we would add a further component to the experiment we sketched in Open Question 1 as described above: we would systematically lesion those brain regions that were activated distally in Open Question 1, and determine whether those lesions blocked not only the emotion state produced by artificial brain stimulation, but by naturalistic exteroceptive stimuli as well. For example, stimulation of the human amygdala evokes strong fMRI activation in the ventromedial prefrontal cortex (Oya et al. 2017). This is what one would expect based on the connectivity between these two structures. It is also borne out by studies in humans that have combined fMRI with lesions. For instance, one fMRI study showed that a task that normally activates the amygdala fails to activate the amygdala in patients who have lesions to the ventromedial prefrontal cortex (Motzkin et al. 2015). A second fMRI study showed that a task that normally activates the medial prefrontal

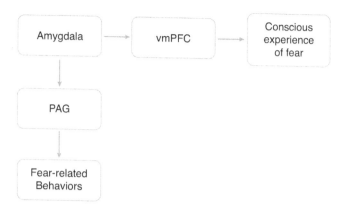

FIGURE 11.1. Hypothesis for how the amygdala produces components of fear. See text for details. vmPFC: ventromedial prefrontal cortex. PAG: periaqueductal gray.

cortex fails to activate the medial prefrontal cortex in patients who have lesions of the amygdala (Hampton et al. 2007). These two studies we just cited suggest that amygdala and medial prefrontal cortex are functionally coupled, and that activation in one influences activation in the other, a finding also borne out by simultaneous single-unit recordings from these structures in monkeys (Morrison et al. 2011).

Returning to our future experiment, if we now stimulate the amygdala optogenetically or electrically in a rodent or human patient, and we find that this induces most or all of the components of an emotion state, what would happen if we additionally lesion the ventromedial prefrontal cortex? Would we lose some or all of the components of the emotion state? This kind of "epistasis" experiment (in which one asks whether inhibition of a "downstream" structure blocks the effect of activating a more "upstream" structure) would be particularly informative in showing us the necessary contribution made by several brain structures. Perhaps the ventromedial prefrontal cortex is particularly important for the conscious experience of an emotion in humans, so that when this region is lesioned, this is the emotion component that would drop out. This would suggest a specific causal graph, as shown in figure 11.1.

According to this scheme, activation of the amygdala (or of certain specific cell types within this structure) is sufficient to produce all the components of fear in a healthy brain. But for all of those components, like the conscious experience of fear, it needs to act through other brain structures. If those are lesioned, we might eliminate that component.

Limitations. As before, there are clear limitations here. Once again, there are technical challenges. If we are working in a mouse, we assume that we would have the precision of optogenetic stimulation and inhibition figured out. But what about the precision of lesioning parts of the prefrontal cortex? If we are using fMRI as the method to visualize the causal effects of amygdala stimulation, for instance, then we would only know that a relatively broad region in the medial prefrontal cortex is activated. But as discussed in chapter 6, there are heterogeneous subsets of neurons within the medial prefrontal cortex, and probably some of these show responses elicited by amygdala stimulation with short latencies, whereas others require longer latencies—and some may not be activated by the amygdala at all. That is, there is a complex circuitry within the medial prefrontal cortex itself, making it complicated to know what to lesion: what, more precisely, do we put into the box for "vmPFC" in figure 11.1? More information about cell types and circuitry in vmPFC will therefore be required in order to make such an experiment meaningful.

A second limitation concerns the idea of mapping the emotion state components onto neural structures in the first place. This exercise might yield a very disjunctive mapping. Depending on how anatomically distributed an emotion state is, and how smeared out in time it is, we might find that lesions of a single structure are generally insufficient to abolish any component of an emotion state, due to redundancy in the system. Perhaps you can only abolish the conscious experience of an emotion state in humans if you lesion many cortical and subcortical regions together (or a larger number of neuronal subpopulations within these structures). That presents a major technical challenge.

The counterpart of this limitation is that lesioning individual structures may cause deficits that do not map easily onto any component of an emotion. Perhaps lesions or reversible inhibition of the ventromedial prefrontal cortex in humans would produce alterations in the reported conscious experience of emotion, but they might also produce other less specific effects that make the experiment harder to interpret. For instance, the patient might say, "Yes, I feel a strong emotion, probably fear, but I'm not really sure and I don't care about it at all." How would we interpret that?

Taken together, the experiments sketched here for Open Questions 1 and 2 are already technically feasible in some aspects but still require further precision and surmounting of methodological and ethical hurdles in their application to humans. But it is likely that the technical limitations will pale in comparison to the conceptual hurdles: how exactly do we assess the emotion state? How do we distinguish it from other central states? How might we distinguish different types of emotion? These questions motivate a second set of open questions that are more conceptual in nature.

Open Question 3. Can We Find States That Satisfy All of Our "Emotion Properties," yet That We Would Not Consider to Be Emotions? From This, Can We Derive Additional Properties That Are Essential to Emotions as Distinct from Other Types of Internal States?

Hypothesis: Our figure 3.2 is incomplete.

The studies we would imagine doing under this heading would require a fair bit of work, since there is a long list of emotion properties. We began listing some back in chapter 3, in figure 3.2. We noted repeatedly that the table is preliminary. So how can we complete it?

One way to do so would be to take a broad inventory, and to collaborate closely with ethologists, to elaborate our initial list. By "broad" we mean broad across species, and broad across central emotion-like states. We would return to the very beginning of our conceptual framework for emotions and survey those states that mediate adaptive responses to recurring environmental themes, at a level of flexibility and complexity that is intermediate between reflexes and deliberative behavior. There is a lot more conceptual work to be done here, just in delineating the environmental themes, and a lot more experimental work to measure behaviors quantitatively and objectively, for example using computer algorithms rather than subjective human observations. Across species, what kinds of defensive behaviors do we observe, and under what circumstances do they appear? Presumably there would be several types, such as those elicited by predators, those elicited by dangerous nonsocial threats (for example, a thunderstorm), and those elicited by aggressive conspecifics. One could imagine a fine-grained typology, since some kinds of threats will require very particular kinds of adaptive responses that only make sense for a particular species in a particular niche. Open Question 3 would require a rather large project that importantly requires careful fieldwork, as well as measurements of internal state variables (hormones, gene expression) and large-scale neuronal population activity. It highlights the cross-species nature that we feel is essential to a science of emotion. This brings us to Open Question 4.

Open Question 4. Are There Species-Specific Emotions?

Hypothesis 4a: There are emotion states, or properties of emotion states, unique to humans.
Hypothesis 4b: There are emotion states, or properties of emotion states, unique to a nonhuman species.

While Hypothesis 4a is commonly encountered, we've never before seen Hypothesis 4b; both seem plausible to us. With respect to Hypothesis 4a, there are some easy candidates, specifically the conscious experience of emotion and the volitional regulation of emotion. There are also particular emotions that may be unique to humans, such as perhaps some social emotions like shame or embarrassment, and emotional feelings like awe. Conversely, there may be emotion states in nonhuman animals—even perhaps in insects—that are not homologous or analogous to any emotion state in humans or higher mammals. This underscores the importance of developing objective criteria for identifying instances of emotional expression in animals that are independent of any anthropocentric or anthropomorphic homologies, but which are based on general properties common to all emotion states (like the ones listed in figure 3.2).

Considering all these different properties, and different emotions, will largely continue the exercise we outlined in Open Question 3; it will ultimately give us a more fine-grained inventory of emotion states and their properties, across a range of species. Once we have this available, we'll need to sit down and decide whether there are principled criteria for allowing some of these as bona fide emotions. Currently, there is no such database available, and the properties of emotions are derived from scattered literature and the intuitions of the experimenters.

Open Question 5: Are There Emotions Unique to Individuals?

Hypothesis: in addition to variability across species, there is variability across individuals.

The above project would raise an important question about individual differences. Instead of asking whether there are emotions unique to a certain species, one might as well ask: are there emotion states that might be unique to a given individual? Is my fear like your fear? Is the brain state that subserves fear in my brain the same as the brain state that subserves fear in your brain? This concern about heterogeneity across individuals is well founded, since it is known that there are large individual differences in the specific stimuli or circumstances that evoke certain emotions in humans. Some people are phobic of spiders, others have them as pets.

You might think that it would help here to return to the experiment described under Open Question 1, and to trigger the emotion, hopefully more reliably, through direct brain activation rather than through sensory stimuli and the possibly idiosyncratic associations these might evoke across different individuals. Might, for example, stimulation of the medial prefrontal cortex evoke an emotion state that is more consistent across people? The answer to this question is not known in humans, but we can get some hints from studies in rodents. As figure 11.2 shows, optogenetic stimulation of the medial prefrontal cortex in rats indeed results in patterns of activation, as visualized with fMRI (exactly the protocol we proposed in Open Question 1), that bear some resemblance to one another. On the other hand, they are far from identical.

The data shown in figure 11.2 raise an important question: how do we decide whether two brain states are instances of the same emotion, or should count as different emotions? Do the subtle differences in fMRI activation that we see across the seven rats in the figure correspond to seven different emotions, or do they correspond to a single emotion

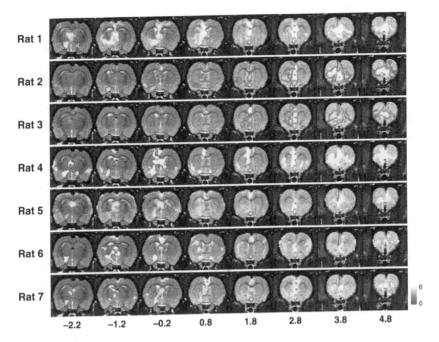

FIGURE 11.2. Combining optogenetics with fMRI. fMRI activation produced by optogenetic stimulation of the infralimbic cortex in seven different rats ("opto-fMRI"). In this study, optogenetic stimulation was achieved in the infralimbic cortex of awake rats, thought to be a homologue of the primate ventromedial prefrontal cortex. The color scale encodes a *t*-statistic of the magnitude of the fMRI activation seen throughout the brain as a result of this optogenetic stimulation. The numbers on the x-axis give the location in the brain at which the shown slice was obtained; each row of brain slices corresponds to different locations in the brain of one of the seven rats. Reproduced with permission from Liang et al. 2015.

with a little variability added in each of the seven rats? Luckily, you will remember that we tackled a problem just like this one in chapter 8, where we discussed representational similarity analysis: we can indeed compare similarity spaces. We have objective ways of comparing how similar or dissimilar the different brain activation patterns shown in figure 11.2 are, and can now complement such low-resolution activity maps with maps at single-cell resolution, using calcium imaging (box 5.2). Once we have a thorough inventory of emotion properties (Open Question 3), we would be in a position to make objective comparisons of how similar or dissimilar emotion states are, as measured by the values of these properties. We could then compare the two similarity spaces—brain activity space and emotion property space—and indeed tackle the questions raised in both Open Questions 4 and 5. We would need a lot of data, but we have the analysis tools available.

The approach here brings us back to what we wrote about testing the efficacy of anxiolytic drugs in chapter 5. We could test how specific drugs affect the pattern of brain activation that characterizes a particular emotion state, and we could test how variable this might be across individuals as well as across species. In short, once we have enough data to construct and relate the similarity spaces, we could decode from the pattern of brain activation what the emotion state is. We could then administer a drug, and see from the change in the pattern of brain activation what change has taken place in the emotion state. Given the relative ease of obtaining fMRI data in human subjects, one could even imagine a future personalized medicine approach, in which this is used to select the most efficacious drug to treat individual human patients.

Open Question 6. Can We Build a Robot That Has Emotions?

Hypothesis. When we understand emotion well enough, we will be able to build robots that have emotions.

The famous physicist and Nobel Laureate Richard Feynman wrote, "What I cannot create, I do not understand," on a blackboard at Caltech. It was found there shortly after his death (figure 11.3). We have proposed that emotions ultimately need to be understood functionally (chapter 2), that this in turn leads to specific properties that emotion states must possess in order to carry out those functions (chapter 3), and that these properties in turn are to be discovered in the brains of animals and people (chapters 4–9). Once we have accumulated knowledge of the functions, properties, and neural implementation of emotions—could we build an artificial system with emotions?

The cyberneticist Valentino Braitenberg wrote a fancifully illustrated book called *Vehicles* in which he sketched how very simple sensors, circuits, and effectors could be combined to produce more and more complex little robots of sorts (Braitenberg 1984). In Braitenberg's imagination, a whole world was soon populated with little machines, all composed of very simple devices that, in their combination, reproduced human behaviors. Indeed, he attributed psychological states including emotions to the more complex vehicles for this reason. One of his intents was to take the magic out of psychological states by showing how they could be engineered from simple components. The components did not have any emotions—but the whole vehicle could, if it had complex enough components.

Certainly, robots have now been built that can fool human observers into attributing emotions to them, by carefully manipulating their posture and sounds (see, for example, the highly expressive robot character BB-8 in the 2015 film *Star Wars: The Force Awakens*). What seems to be a more difficult challenge is to build an artificial system in which emotion states are brought into play on some occasions, and not others, to influence the system's behavior, decision-making, and other processes that influence the system's "survival." Ultimately, whether an emotion system has been successfully instantiated in an artificial entity should be determined by the utility or adaptive value that such a system provides to the entity itself—not by whether it "looks" or "sounds" like an emotion state to a human observer. That evaluation will, in turn, require an understanding of the ethology and psychology of the entity, and a willingness to study that entity's behavior in its ecological niche on its own terms—very much the approach that we advocate in the study of emotion states in animals. There is of course one further wrinkle on this: insofar as emotion states function in the service of social interactions, it may indeed be a goal to design robots to mimic human-like emotional behaviors, because this will serve the function of social interaction with humans. This is currently in fact the main goal of designing "emotional" robots.

Many aspects in the design of complex artificial control systems may be similar to the challenges that natural history faced in evolving natural control systems in animals. We would therefore think it very plausible that a sufficiently complex artificial system, operating in a sufficiently complex and open-ended environment (including the social environment), would also require for its control certain system states that are functionally equivalent to emotions. As we noted in chapter 3 already, emotion states are in fact distributed not only spatially but also temporally, including fast and slow processing aspects. Much the same is true of artificial control systems, such as the power grid or data transfer

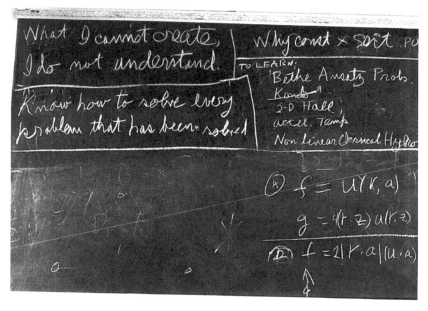

FIGURE 11.3. Richard Feynman's blackboard at the time of his death. The quote, at the upper left, is often heard in engineering and the physical sciences, and suggests the importance of finding causal mechanisms. Courtesy of the Archives, California Institute of Technology.

over the internet. Some of the dichotomies that we listed in figure 3.6 will apply to engineered systems as well.

We are with Feynman in believing that the ultimate test of understanding emotions will come when we can engineer them. Although it is currently impossible to create emotional robots (that is, robots with internal emotion states, rather than automata that can mimic the expression of an emotion, like a smile on the face of a robotic monkey), we believe that the eventual design of robots that are truly autonomous and that can function flexibly in the real world will in fact require building in emotions. We also believe that the only way we will figure out how to build emotions into robots is through painstakingly discovering how emotions are instantiated in the brains of animals and people.

GLOSSARY

Acetylcholine (ACh): a neurotransmitter.

Amygdala: a collection of nuclei in the medial temporal lobe involved in emotion and emotional learning.

Antennal lobe: a structure in the insect brain that receives input from primary olfactory sensory neurons, analogous to the olfactory bulb in vertebrates.

Anterior cingulate cortex (ACC): a part of the prefrontal cortex (PFC, see below) involved in emotion and other functions, such as attention.

Aplysia californica: an invertebrate sea slug whose synaptic mechanisms for learning and memory are particularly well understood.

Appraisal: a feature of psychological emotion theories describing how emotions are induced through a complex series of evaluations.

Arousal: a state characterized by increased behavioral activity and sensitivity to sensory stimuli. Arousal states are graded, and related to the emotion properties of intensity or scalability. There may be different types of arousal associated with different emotion states. See figure 3.2 and chapter 5.

Arthropods: phylum comprising invertebrate animals with articulated exoskeletons, including insects, spiders, and crustaceans.

Automaticity: a property of emotions describing their priority over behavioral control, a common component of dual process models. See figures 3.2 and 3.6.

Autonomic nervous system (ANS): in distinction to the somatic nervous system, which controls skeletal muscles, the part of the central and peripheral nervous system that controls smooth muscles (e.g., the pupil of the eye) and hormones (e.g., cortisol). The ANS comprises the sympathetic and parasympathetic nervous systems.

Axon: that part of a neuron from which electrical potentials convey information to other neurons, often over a long distance.

Basolateral amygdala (BLA): a nucleus in the amygdala (see above) that has extensive connections with cortex.

Bed nucleus of the stria terminals (BNST): a collection of nuclei closely interconnected with the amygdala (see above) and involved in emotion, especially fear and anxiety.

BOLD signal: <u>B</u>lood <u>O</u>xygenation <u>L</u>evel-<u>D</u>ependent signal detected by fMRI (see below) a surrogate measure of neuronal activity.

Building blocks of emotion: properties of an emotion that are more essential, and more basic, also called "emotion primitives." Simpler organisms show building blocks of emotion. Features (see below), on the other hand, are more elaborated, derived, and variable properties of emotions. To use an automotive analogy, wheels are a building block, while air conditioning is a feature.

C57BL/6J mice: an inbred strain of mice used in research. It provides a uniform and inbred genetic background against which to more clearly see the effects of mutations.

***Caenorhabditis elegans*:** a genetically accessible nematode worm whose nervous system has only about three hundred neurons, and whose complete connectivity (connectome) has been mapped.

Calcium imaging: visualizing neuronal activity with single-cell resolution by microscopically measuring increases in fluorescence intensity that occur when calcium levels increase as a consequence of action potential firing, due to the binding of calcium to a genetically encoded fluorescence indicator such as GCaMP (see boxes 5.1–5.3).

Cell type: a particular type of neuron that is distinguishable from other types of neuron by multiple criteria, including cellular morphology, projections, gene expression profile and biophysical properties (electrical excitability). The total number of different cell types in the brain is not yet known. The visual cortex is thought to contain fifty to one hundred different cell types, half of them excitatory and half inhibitory.

Central nucleus of the amygdala (CeA, CEA, CE): a nucleus in the amygdala (see above) that has extensive connections with brainstem and hypothalamus. It consists of further subdivisions, such as the lateral subdivision of the CeA (CeL) and the medial subdivision of the CeA (CeM).

Cephalopods: a taxonomic class comprised of invertebrates with legs attached to their heads, including octopuses, cuttlefish, and squid.

c-fos: an immediate early gene (IEG, see below) whose mRNA or protein products are used as a reporter of neuronal activity. See box 6.3.

Channelrhodopsin-2: A light-activated cation channel (opsin) from *Chlamydomonas reinhardii* used to activate neurons in optogenetic experiments (see below)

Cognitive bias: a term used to describe an influence of emotion states on cognitive processes, such as decision-making. Cognitive bias can be negative (e.g., "pessimism") or positive (e.g., "optimism"). Cognitive bias has been used to study emotion states in honeybees and several other animal species.

Concepts: what we use to think about things. Thinking about emotions requires having a concept of emotion.

Conditioned stimulus (CS): in associative learning, a stimulus that has acquired the ability to cause a response through association with another stimulus (the unconditioned stimulus, US). See figure 6.1.

Connectome: a "wiring diagram" illustrating the complete set of connections between all the neurons in an animal's nervous system. Complete connectomes are available currently only for *C. elegans* (see above) and *Drosophila* larvae.

Construct validity: the adequacy with which a dependent measure (e.g., task performance) can be used to infer a latent variable (e.g., a central emotion state). See box 8.1.

Cortex: a specialized, laminated structure of the brain found only in mammals. It is greatly enlarged in primates and especially humans, and the part of the brain visible when you look at it its outside surface.

Cre: a type of enzyme called a "recombinase," which can be used to rearrange pieces of DNA in a way that promotes or restricts expression of foreign genes. Genetically targeted expression of Cre in specific neuronal cell types allows optogenetic or pharmacogenetic (see below) manipulations of those neurons.

Deep-brain stimulation (DBS): electrical stimulation of the brain through implanted electrodes. Primarily used to treat movement disorders like Parkinson's disease, it is also being explored to treat mood disorders like depression.

Demand characteristics: behaviors evoked in a human subject experiment that are due to the external demands of the study, for instance the subjects knowing what the experimenter would like to see or hypothesizes. This is a major confound in human studies.

Dendrite: part of a neuron that receives inputs from other neurons. It is now known that dendrites perform complex computations, and fire action potentials.

Domain specificity: specialized processing for a particular domain of stimuli, like faces, language, or predators. See box 3.1.

Dopamine (DA): a kind of neuromodulator called a biogenic amine, involved in reward, arousal, and emotion.

Dorsomedial prefrontal cortex (dmPFC): part of the prefrontal cortex (PFC, see below) involved in causal inference and the attribution of emotion.

DREADDs: Designer receptors exclusively activated by designer drugs. Engineered G protein-coupled receptors (GPCRs) that can be expressed in specific neurons and activated by synthetic drugs, and which are not activated by any endogenous brain signaling molecules. Some DREADDs lead to neuronal activation, while others lead to inhibition. In this way, neuronal activity can be increased or decreased by systemic administration of the drug (which crosses the blood-brain barrier).

Drosophila melanogaster: commonly called the fruit fly, more correctly the vinegar fly, this insect is a model organism for understanding behavior and emotion through genetic mutations and genetic manipulations of neuronal activity.

Elevated plus maze (EPM): a behavioral task used to measure anxiety in rodents. See figure 6.6.

Emotion primitives: see "building blocks of emotion."

Esr1: the type-1 estrogen receptor, which is expressed in specific populations of neurons, providing genetic access to those cells for the identification and manipulation of activity, through methods from molecular biology.

Features of emotion: see "building blocks of emotion."

Feelings: subjective, conscious experiences of emotions and other states.

fMRI: functional magnetic resonance imaging. A method to measure brain activity reflected by BOLD signals (see above) using an MRI scanner, most commonly used to investigate emotion and cognition in humans.

GABA: gamma aminobutyric acid, an inhibitory neurotransmitter.

Generalization: a property of emotions describing how many stimuli can cause one emotion, which in turn can cause many behaviors ("Fan-in, fan-out" architecture). See figure 3.2.

Generalized anxiety disorder (GAD): a mood disorder featuring pervasive anxiety; distinct from other anxiety disorders, such as phobias, which are specific to certain stimuli.

Germline modifications: engineered modifications of DNA in an organism that are transmitted to offspring.

Glutamate: an excitatory neurotransmitter.

Halorhodopsin: A microbial light-activated chloride pump (opsin), used to inhibit neuronal activity in optogenetic experiments (see below).

Homunculus: in philosophy and psychology, the fallacious idea that our brain contains the equivalent of a little person who can perceive all mental activity of which we are conscious. In neuroscience, the term is used to refer to sensory or motor representations whose anatomical organization bears resemblance to the organization of parts of the body, as is found in somatosensory cortex.

Hypothalamus: a deep subcortical collection of nuclei, located below the thalamus, that control innate, motivated behaviors such as mating, fighting, and predator defense as well as homeostatic functions (e.g., eating and drinking).

IEG: immediate early gene. See c-fos (above).

Infralimbic cortex (IL): cortex in the frontal lobe of the rodent brain, which together with Prelimbic cortex (PrL, see below) forms the medial prefrontal cortex (mPFC), thought to be homologous to ventromedial prefrontal cortex (see below) in primates.

Innate: determined to develop from birth. Also termed instinctive, and distinct from genetic, acquired, and congenital. The capacity for language is innate in humans, even though it involves both genetic and acquired factors and is not congenital but emerges over development.

Instructive: in distinction to permissive, a causal factor that carries information. In making your car drive, how hard you press the gas pedal is instructive, whereas having any gas in your tank is permissive.

Intercalated cell mass (ic, ITC): a somewhat distributed nucleus in the amygdala.

Interneuron: a type of neuron located within a particular brain region or structure that makes local connections with other neurons in that structure. Interneurons are typically inhibitory, but can be excitatory as well.

Interoception: the perception of the internal state of the body, usually nonconsciously.

Kenyon cell: a type of neuron in the insect mushroom body (see below) that represents a specific odor, via input from the antennal lobe (see above).

Lateral amygdala (LA): lateral amygdalar nucleus, sometimes used interchangeably with BLA, which combines the LA and Basal Amygdala.

MAD states: motivation, arousal, and drive states. See chapter 5.

Medial amygdala (MeA): medial amygdala, an amygdala nucleus receiving input from the accessory olfactory system and projecting to the BNST (see above) and hypothalamus.

Metacognition: cognition about cognition, usually referring to "higher-order" control and awareness of cognitive processes. The clearest example of a metacognitive ability is the ability to provide confidence judgments about one's own task performances.

Motivation: an internal state, related to (but distinct from) emotion, which drives an animal to engage in a goal-directed behavior. Motivational states can be triggered both by interoceptive stimuli (e.g., water or caloric deprivation) or exteroceptive stimuli (e.g., money).

MPOA: medial preoptic area, a region of the hypothalamus (see above) involved in mating-related behaviors.

MRI: magnetic resonance imaging. See figure 1.1 for an example and chapter 8 for an explanation of the method.

Mushroom body: a structure in the insect brain involved in learning and memory.

Natural kind: in philosophy, a category that is thought to exist objectively in nature (like the chemical elements) rather than by social convention (like different monetary currencies).

NE: norepinephrine, a neuromodulator related to adrenaline.

Neuromodulator: a type of chemical signal in the brain that acts to modulate neuronal activity on a slower time-scale (seconds, minutes) than fast transmitters (milliseconds). Neuromodulators include neuropeptides (see below) and small molecules such as dopamine, norepinephrine, or acetylcholine.

Neuropeptide: a type of neurotransmitter consisting of small protein-like molecules that are encoded by specific genes.

Neurotransmitter: a molecule released by one neuron in order to provide chemical communication with other neurons through synapses.

Nucleus: neuroanatomical term referring to a group of densely packed neuronal cell bodies, as distinct from a laminated structure (e.g., cortex or hippocampus) where neurons are arranged in layers. Not to be confused with the nucleus of a cell, which contains its chromosomes. Examples of neuroanatomical nuclei are the amygdala and the hypothalamus.

Nucleus accumbens (NAc, NAcc): a nucleus of the basal ganglia involved in reward processing.

Open field test (OFT): a behavioral task used to measure anxiety in rodents. See figure 6.6.

Opsins: Light-activated ion channels or pumps used in optogenetic experiments (see below), such as channelrhodopsin-2 and halorhodopsin (see above).

Optogenetics: genetic modification of neurons to permit regulation of their activity by light, through the expression of opsins (see above) (see boxes 5.1–5.3).

Optomotor response: visually guided behavior typically studied in insects, in which the animal alters its locomotor trajectory to maintain a moving optical stimulus (e.g., vertical bar) in the center of its visual field.

Pavlovian fear conditioning: a form of associative learning, see figure 6.1.

Periaqueductal gray (PAG): a brainstem region involved in the expression of emotional behaviors.

Permissive: in distinction to instructive (see above).

Persistence: a property of emotions related to how long they last. See figure 3.2.

Pharmacogenetics: genetic modification of neurons to permit regulation of their activity by specific chemicals or drugs (e.g., DREADDS, see above), also called "chemogenetics (see boxes 5.1–5.3).

PR: progesterone receptor. A type of steroid hormone receptor whose expression can be used to distinguish neuronal cell types.

Prefrontal cortex (PFC): the most anterior portion of cortex, enlarged in primates, and often subdivided further. mPFC, vmPFC, dmPFC, and dlPFC refer to medial, ventromedial, dorsomedial, and dorsolateral prefrontal cortex, respectively. Involved in many cognitive functions including emotions.

Prelimbic cortex (PrL): a part of the rodent mPFC, together with the infralimbic cortex (IL, see above). PrL and IL exert opposite-direction influences on the amygdala.

Projection neuron: a type of neuron whose axons make long-range projections out of the structure in which it is located.

Proper function: the function for which a biological system or organ evolved. See chapter 2.

Prozac: an SSRI drug (see below) used to treat depression and anxiety.

Post-traumatic stress disorder (PTSD): an anxiety disorder in which traumatic experiences cause emotional learning that bears some resemblance to Pavlovian fear conditioning.

Psychophysiology: measures of autonomic nervous system function, such as heart rate and skin-conductance. Commonly used to measure emotions in human psychological studies.

Regulation of emotion: the ability to control your emotions. See box 3.4.

Representational similarity analysis (RSA): a method to quantify similarity between patterns, commonly applied to patterns of brain activation measured with fMRI. See figure 8.4.

Scalability: a property of emotion states describing their magnitude or intensity, see figure 3.2.

Serotonin: a biogenic amine neurotransmitter (like Dopamine).

Sf1/Nr5a1: a genetic marker for a specific subpopulation of neurons in VMHdm/c (see below).

SSRI: selective serotonin reuptake inhibitor. A class of drugs, such as Prozac, that are used to treat mood disorders.

Subcortical: pertaining to structures below the level of cortex, such as the amygdala, hypothalamus, and brainstem.

Superior temporal sulcus (STS): together with other cortex in the temporal lobe, this region processes emotional stimuli such as faces and body postures when we observe another person's behavior.

Supervenience: in philosophy, A supervenes on B if and only if any change we can detect in A requires a change in B (but not conversely). Mental states are usually thought to supervene on brain states.

Synapse: a specialized region where two neurons communicate with one another through neurotransmitters.

Token identity: in philosophy, an identity relation between individual events (tokens). The claim that a particular person's state of fear on a particular occasion is identical with a particular brain state, is an example of token identity.

Type identity: by contrast to token identity, this is a relation between categories of events. The claim that an emotion category, like fear, in general across people and across different instances, is identical to type of brain state, is an example of type identity. Type identity is much more tricky to establish then token identity.

Unconditioned stimulus (US): in associative learning, a stimulus that causes a response (the unconditioned response, UR) without learning. See figure 6.1.

Valence: a property of emotions describing the dimension of pleasantness/unpleasantness. See figure 3.2 and chapter 10.

Ventral tegmental area: a subcortical region enriched in dopamine neurons thought to represent reward, or reward prediction error, and which projects to the nucleus accumbens (see above).

Ventromedial hypothalamus (VMH): a region of the hypothalamus (see above) involved in defensive behaviors, aggression, and mating. It is further subdivided into dorsomedial, ventrolateral, and central regions (VMHdm, VMHvl, and VMHc).

REFERENCES

Adolphs, R. (2013). "The biology of fear." *Current Biology* 23: R79–R93.

Adolphs, R. (2016). "Human lesion studies in the 21st century." *Neuron* 90 (6): 1151–53.

Adolphs, R. (2017). "How should neuroscience study emotions? By distinguishing emotion states, concepts, and experiences." *Social Cognitive and Affective Neuroscience* 12: 24–31.

Adolphs, R., and D. Andler. (2018, in press). "Investigating emotions as functional states distinct from feelings." *Emotion Review.*

Adolphs, R., F. Gosselin, T. W. Buchanan, D. Tranel, P. Schyns, and A. R. Damasio. (2005a). "A mechanism for impaired fear recognition after amygdala damage." *Nature* 433: 68–72.

Adolphs, R., D. Tranel, M. Koenigs, and A. Damasio. (2005b). "Preferring one taste over another without recognizing either." *Nature Neuroscience* 8: 860–61.

Anderson, A. K., K. Christoff, I. Stappen, D. Panitz, D. G. Ghahremani, G. Glover, J. D. E. Gabrieli, and N. Sobel. (2003). "Dissociated neural representations of intensity and valence in human olfaction." *Nature Neuroscience* 6: 196–202.

Anderson, D. J. (2016). "Circuit models linking internal states and social behavior in flies and mice." *Nature Reviews Neuroscience* 17: 692–704.

Anderson, D. J., and R. Adolphs. (2014). "A framework for studying emotions across species." *Cell* 157 (1): 187–200.

Anderson, S. W., J. Barrash, A. Bechara, and D. Tranel. (2006). "Impairments of emotion and real-world complex behavior following childhood- or adult-onset damage to ventromedial prefrontal cortex." *Journal of the International Neuropsychological Society* 12: 224–35.

Aryal, M. F., S. Ho, J. Stewart, J. C. Norman, N. A. Tan, T. L. Eisaka, and A. Claridge-Chang. (2016). "Ancient anxiety pathways influence *Drosophila* defense behaviors." *Current Biology* 26: 981–86.

Aso, Y., D. Sitaraman, T. Ichinose, K. R. Kaun, K. Vogt, G. Belliart-Guérin, P.-Y. Plaçais, A. A. Robie, N. Yamagata, C. Schnaitmann, et al. (2014). "Mushroom body output neurons encode valence and guide memory-based action selection in *Drosophila.*" *eLife* 3: e04580.

Bach, D. R., and P. Dayan. (2017). "Algorithms for survival: A comparative perspective on emotions." *Nature Reviews Neuroscience* 18: 311–19.

Bateson, M., S. Desire, S. E. Gartside, and G. A. Wright. (2011). "Agitated honeybees exhibit pessimistic cognitive biases." *Current Biology* 21: 1070–73.

Bekoff, M. (2007). *The Emotional Lives of Animals.* New York: New World Library.

Berns, G. (2017). *What It's Like to Be a Dog.: And Other Adventures in Animal Neuroscience.* New York: Basic Books.

Berns, G. S., A. M. Brooks, and M. Spivak. (2012). "Functional MRI in awake unrestrained dogs." *PLoS One* 7: e38027.

Berridge, K. C. (2003). "Pleasures of the brain." *Brain and Cognition* 52: 106–28.

Berridge, K. C. (2004). "Motivation concepts in behavioral neuroscience." *Physiology and Behavior* 81 (2): 179–209.

Blood, A. J., and R. J. Zatorre. (2001). "Intensely pleasurable responses to music correlate with activity in brain regions implicated in reward and emotion." *PNAS* 98: 11818–23.

Braitenberg, V. (1984). *Vehicles: Experiments in Synthetic Psychology.* Cambridge, MA: MIT Press.

Bromberg-Martin, E. S., M. Matsumoto, and O. Hikosaka. (2011). "Dopamine in motivational control: Rewarding, aversive, and alerting." *Neuron* 68: 815–34.

Brown, M. E. (2010). *How I Killed Pluto and Why It Had It Coming.* eBook, ISBN 0-385-53108-7.

Cacioppo, J. T., L. G. Tassinary, and G. G. Berntson. (2007). *Handbook of Psychophysiology,* 3rd ed. Cambridge: Cambridge University Press.

Caldara, R. (2017). "Culture reveals a flexible system for face processing." *Current Directions in Psychological Science* 26: 249–55.

Calder, A. J., A. D. Lawrence, and A. W. Young. (2001). "Neuropsychology of fear and loathing." *Nature Reviews Neuroscience* 2: 352–63.

Canteras, N. S. (2002). "The medial hypothalamic defensive system: Hodological organization and functional implications." *Pharmacology, Biochemistry, and Behavior* 71: 481–91.

Card, G., and M. H. Dickinson. (2008). "Visually mediated motor planning in the escape response of *Drosophila.*" *Current Biology* 18: 1300–1307.

Cohn, R., I. Morantte, and V. Ruta. (2015). "Coordinated and compartmentalized neuromodulation shapes sensory processing in *Drosophila.*" *Cell* 163: 1742–55.

Cowey, A., and P. Stoerig. (1995). "Blindsight in monkeys." *Nature* 373: 247–49.

Craig, A. D. (2002). "How do you feel? Interoception: The sense of the physiological condition of the body." *Nature Reviews Neuroscience* 3 (8): 655–66.

Crick, F. H. C. (1988). *What Mad Pursuit: A Personal View of Scientific Discovery.* New York: Basic Books.

Damasio, A. R. (1995). "Toward a neurobiology of emotion and feeling: Operational concepts and hypotheses." *Neuroscientist* 1: 19–25.

Damasio, A. R. (1999). *The Feeling of What Happens: Body and Emotion in the Making of Consciousness.* New York: Harcourt Brace.

Damasio, A. R. (2003). *Looking for Spinoza: Joy, Sorrow, and the Feeling Brain.* Orlando, FL: Harcourt.

Damasio, A. R., and G. B. Carvalho. (2013). "The nature of feelings: Evolutionary and neurobiological origins." *Nature Reviews Neuroscience* 14 (2): 143–52.

Damasio, A. R., T. Grabowski, et al. (2000). "Feeling emotions: Subcortical and cortical brain activity during the experience of self-generated emotions." *Nature Neuroscience* 3: 1049–56.

Damasio, H., T. Grabowski, R. Frank, A. M. Galaburda, and A. R. Damasio. (1994). "The return of Phineas Gage: Clues about the brain from the skull of a famous patient." *Science* 264: 1102–4.

Darwin, C. (1872/1965). *The Expression of the Emotions in Man and Animals.* Chicago: University of Chicago Press.

Davis, M. (1992). "The role of the amygdala in fear and anxiety." *Annual Review of Neuroscience* 15: 353–75.

De Gelder, B., M. Tamietto, G. van Boxtel, R. Goebel, A. Sahraie, J. van den Stock, B. Stienen, L. Weiskrantz, and A. Pegna. (2008). "Intact navigation skills after bilateral loss of striate cortex." *Current Biology* 18: 10.1016/j.cub.2008.1011.1002.

Dejjani, B.-P., P. Damier, et al. (1999). "Transient acute depression induced by high-frequency deep-brain stimulation." *New England Journal of Medicine* 340: 1476–80.

Dilks, D. D., P. Cook, S. K. Weiller, H. P. Berns, M. Spivak, and G. S. Berns. (2015). "Aware fMRI reveals a specialized region in dog temporal cortex for face processing." *PeerJ* 3: e1115.

Dixon, M. L., R. Thiruchselvam, R. Todd, and K. Christoff. (2017). "Emotion and the prefrontal cortex: An integrative review." *Psychological Bulletin* 143: 1033–81.

Dolan, R. J. (2002). "Emotion, cognition, and behavior." *Science* 298 (5596): 1191–94.

Dubois, J., and R. Adolphs. (2015). "Neuropsychology: How many emotions are there? (Dispatch)." *Current Biology* 25: R669–R672.

Dubois, J., and R. Adolphs. (2016). "Building a science of individual differences from fMRI." *Trends in Cognitive Sciences* 20: 425–43.

Dubois, J., H. Oya, J. M. Tyszka, M. Howard, F. Eberhardt, and R. Adolphs. (2018, in press). "Causal mapping of emotion networks in the human brain: A framework and preliminary findings." *Neuropsychologia*.

Eberhardt, F. (2017). "Introduction to the foundations of causal discovery." *International Journal of Data Science and Analytics* 3: 81–91.

Ekman, P. (1972). "Universals and cultural differences in facial expressions of emotion." In *Nebraska Symposium on Motivation, 1971*, edited by J. Cole, 207–83. Lincoln: University of Nebraska Press.

Engle, R. W., and T. R. Zentall. (2016). Special Issue on Cognition in Dogs. *Current Directions in Psychological Science* 25 (5).

Ekman, P. (1994). "Strong evidence for universals in facial expressions: A reply to Russell's mistaken critique." *Psychological Bulletin* 115: 268–87.

Fanselow, M. (1980). "Conditional and unconditional components of post-shock freezing in rats." *Pavlovian Journal of Biological Sciences* 15: 177–82.

Fanselow, M. (1984). "What is conditioned fear?" *Trends in Neurosciences* 7: 460–62.

Fanselow, M. S., and Z. T. Pennington (2018). "A return to the psychiatric dark ages with a two-system framework for fear." *Behavior Research and Therapy* 100: 24–29.

Fernandez-Dols, J.-M., and J. A. Russell. (2017). *The Science of Facial Expression*. New York: Oxford University Press.

Feinstein, J. S., R. Adolphs, A. Damasio, and D. Tranel. (2011). "The human amygdala and the induction and experience of fear." *Current Biology* 21: 34–38.

Feinstein, J. S., R. Adolphs, and D. Tranel. (2016). "A tale of survival from the world of patient S.M." In *Living without an Amygdala*, edited by D. G. Amaral and R. Adolphs. New York: Guilford Press.

Feinstein, J. S., C. Buzza, R. Hurlemann, R. L. Follmer, N. S. Dahdaleh, W. H. Coryell, M. J. Welsh, D. Tranel, and J. A. Wemmie. (2013). "Fear and panic in humans with bilateral amygdala damage." *Nature Neuroscience* 16: 270–72.

Feinstein, J. S., M. C. Duff, and D. Tranel. (2010). "Sustained experience of emotion after loss of memory in patients with amnesia." *PNAS* 107: 7674–79.

Feldman Barrett, L. (2015). "What emotions are (and aren't)." *New York Times*.

Feldman Barrett, L. (2017a). *How Emotions Are Made: The Secret Life of the Brain*. New York: Houghton Mifflin Harcourt.

Feldman Barrett, L. (2017b). "The theory of constructed emotion: An active inference account of interoception and categorization." *Social Cognitive and Affective Neuroscience* 12: 17–23.

Feldman Barrett, L., B. Mesquita, K. N. Ochsner, and J. J. Gross. (2007). "The experience of emotion." *Annual Review of Psychology* 58: 373–403.

Feldman Barrett, L., and A. B. Satpute. (2013). "Large-scale brain networks in affective and social neuroscience: Towards an integrative functional architecture of the brain." *Current Opinion in Neurobiology* 23: 361–72.

Filippi, P., J. V. Congdon, J. Hoang, D. L. Bowling, S. A. Reber, A. Pasukonis, M. Hoeschele, S. Ocklenburg, C. deBoer, C. B. Sturdy, et al. (2017). "Humans recognize emotional arousal in vocalizations across all classes of terrestrial vertebrates: Evidence for acoustic universals." *Proceedings of the Royal Society of London B: Biological Sciences* 284: doi: 10.1098/rspb.2017.0990.

Fisher, R. A. (1990). *Statistical Methods: Experimental Design and Scientific Inference*. New York: Oxford University Press.

Flavell, S. W., N. Pokala, E. Z. Macosko, D. R. Albrecht, J. Larsch, and C. I. Bargmann. (2013). "Serotonin and the neuropeptide PDF initiate and extend opposing behavioral states in *C. elegans*." *Cell* 154: 1023–35.

Fodor, J. A. (1983). *The Modularity of Mind: An Essay on Faculty Psychology*. Cambridge, MA: MIT Press.

Fossat, P., J. Bacqué-Cazenave, P. De Deurwaerdère, J.-P. Delbecque, and D. Cattaert. (2014). "Comparative behavior: Anxiety-like behavior in crayfish is controlled by serotonin." *Science* 344: 1293–97.

Fridlund, A. J. (1994). *Human Facial Expression.* New York: Academic Press.

Fried, I., U. Rutishauser, M. Cerf, and G. Kreiman. (2014). *Single-Neuron Studies of the Human Brain: Probing Cognition.* Cambridge, MA: MIT Press.

Fried, I., C. L. Wilson, K. A. MacDonald, and E. J. Behnke. (1998). "Electric current stimulates laughter." *Nature* 391: 650.

Gendron, M., D. Robertson, J. M. van der Vyver, and L. Feldman Barrett. (2014). "Perceptions of emotion from facial expressions are not culturally universal: Evidence from a remote culture." *Emotion* 14: 251–62.

Gibson, W. T., C. R. Gonzalez, C. Fernandez, L. Ramasamy, T. Tabachnik, R. R. Du, P. D. Felsen, M. R. Maire, P. Perona, and D. J. Anderson. (2015). "Behavioral responses to a repetitive visual threat stimulus express a persistent state of defensive arousal in *Drosophila.*" *Current Biology* 25: 1401–15.

Goldman, A. I., and C. S. Sripada. (2005). "Simulationist models of face-based emotion recognition." *Cognition* 94 (3): 193–213.

Grewe, B. F., J. Gründemann, L. J. Kitch, J. A. Lecoq, J. G. Parker, J. D. Marshall, M. C. Larkin, P. E. Jercog, F. Grenier, J. Z. Li, A. Lüthim, and M. J. Schnitzer. (2017). "Neural ensemble dynamics underlying a long-term associative memory." *Nature* 543: 670–75.

Griebel, G., and F. Holsboer. (2012). "Neuropeptide receptor ligands as drugs for psychiatric diseases: The end of the beginning?" *Nature Reviews Drug Discovery* 11: 1–17.

Griffiths, P. E. (1997). *What Emotions Really Are.* Chicago: University of Chicago Press.

Gross, C. T., and N. S. Canteras. (2012). "The many paths to fear." *Nature Reviews Neuroscience* 13: 651–58.

Gross, J. J. (2015). "Emotion regulation: Current status and future prospects." *Psychological Inquiry* 26: 1–26.

Gross, J. J., and R. W. Levenson. (1995). "Emotion elicitation using films." *Cognition and Emotion* 9: 87–107.

Guillory, S. A., and K. A. Bujarski. (2014). "Exploring emotions using invasive methods: A review of 60 years of human intracranial electrophysiology." *Social Cognitive and Affective Neuroscience* 9: 1880–89.

Hampton, A., R. Adolphs, J. M. Tyszka, and J. O'Doherty. (2007). "Contributions of the amygdala to reward expectancy and choice signals in human prefrontal cortex." *Neuron* 55: 545–55.

Han, S., M. T. Soleiman, M. E. Soden, L. S. Zweifel, and R. D. Palmiter. (2015). "Elucidating an affective pain circuit that creates a threat memory." *Cell* 162: 363–74.

Heisenberg, M. (2003). "Mushroom body memoir: From maps to models." *Nature Reviews Neuroscience* 4: 266–75.

Hess, W. R., and M. Brügger. (1943). "Das subkortikale Zentrum der affecktiven Abwehrreaktion." *Helvetica Physiologica et Pharmacologica Acta* 1: 33–52.

Hornak, J., J. Bramham, E. T. Rolls, R. G. Morris, J. O'Doherty, P. R. Bullock, and C. E. Polkey. (2003). "Changes in emotion after circumscribed surgical lesions of the orbitofrontal and cingulate cortices." *Brain* 126: 1691–712.

Hornak, J., E. T. Rolls, and D. Wade. (1996). "Face and voice expression identification in patients with emotional and behavioral changes following ventral frontal lobe damage." *Neuropsychologia* 34: 247–61.

Hutcherson, C., B. Bushong, and A. Rangel. (2015). "A neurocomputational model of altruistic choice and its implications." *Neuron* 87: 451–62.

James, W. (1884). "What is an emotion?" *Mind* 9: 188–205.

Janak, P. H., and K. M. Tye. (2015). "From circuits to behaviour in the amygdala." *Nature* 517: 284–92.

Jiang, Y., P. Costello, F. Fang, M. Huang, and S. He. (2006). "A gender and sexual orientation-dependent spatial attentional effects of invisible images." *PNAS* 103: 17048–52.

Jin, J., J. A. Gottfried, and A. Mohanty. (2015). "Human amygdala represents the complete spectrum of subjective valence." *Journal of Neuroscience* 35: 15145–56.

Kahneman, D. (2011). *Thinking, Fast and Slow*. New York: Farrar, Straus, and Giroux.

Kandel, Eric R., James H. Schwartz, Thomas M. Jessell, Steven A. Siegelbaum, and A. J. Hudspeth, eds. (2013). *Principles of Neural Science*. 5th ed. New York: McGraw-Hill Medical.

Kanwisher, N. (2000). "Domain specificity in face perception." *Nature Neuroscience* 3: 759–63.

Kashdan, T. B., L. Feldman Barrett, and P. E. McKnight. (2015). "Unpacking emotion differentiation: Transforming unpleasant experience by perceiving distinctions in negativity." *Current Directions in Psychological Science* 24: 10–16.

Kennedy, D., and R. Adolphs. (2012). "Perception of emotions from facial expressions in high-functioning adults with autism." *Neuropsychologia* 50: 3313–19.

Khalsa, S. S., D. Rudrauf, J. S. Feinstein, and D. Tranel. (2010). "The pathways of interoceptive awareness." *Nature Neuroscience* 12: 1494–96.

Killingsworth, M. A., and D. T. Gilbert. (2010). "A wandering mind is an unhappy mind." *Science* 330: 932.

Kim, J., X. Zhang, S. Muralidhar, S. A. LeBlanc, and S. Tonegawa. (2017). "Basolateral to central amygdala neural circuits for appetitive behaviors." *Neuron* 93: 1464–79.

Koch, C. (2004). *The Quest for Consciousness*. Englewood, CO: Roberts.

Koenigs, M., and D. Tranel. (2007). "Irrational economic decision-making after ventromedial prefrontal damage: Evidence from the ultimatum game." *Journal of Neuroscience* 27: 951–56.

Koenigs, M., L. Young, R. Adolphs, D. Tranel, F. Cushman, M. Hauser, and A. Damasio. (2007). "Damage to the prefrontal cortex increases utilitarian moral judgments." *Nature* 446: 908–11.

Kragel, P. A., A. R. Knodt, A. R. Hariri, and K. S. LaBar. (2016). "Decoding spontaneous emotional states in the human brain." *PloS Biology* 14: e2000106.

Kriegeskorte, N., and R. A. Kievit. (2013). "Representational geometry: Integrating cognition, computation, and the brain." *Trends in Cognitive Sciences* 8: 401–12.

Kunwar, P. S., M. Zelikowsky, R. Remedios, H. Cai, M. Yilmaz, M. Meister, and D. J. Anderson. (2015). "Ventromedial hypothalamic neurons control a defensive emotion state." *eLife* 4.

Lammel, S., B. K. Lim, C. Ran, K. W. Huang, M. J. Betley, K. M. Tye, K. Deisseroth, and R. C. Malenka. (2012). "Input-specific control of reward and aversion in the ventral tegmental area." *Nature* 491: 212–17.

Lawrence, P. (1992). *The Making of a Fly: The Genetics of Animal Design*. Oxford: Blackwell Scientific Publications.

Lazarus, R. S. (1991). *Emotion and Adaptation*. New York: Oxford University Press.

Lebestky, T., J.-S. C. Chang, H. Dankert, L. Zelnik, Y.-C. Kim, K.-A. Han, F. W. Wolf, P. Perona, and D. J. Anderson. (2009). "Two different forms of arousal in *Drosophila* are oppositely regulated by the dopamine D1 receptor ortholog DopR via distinct neural circuits." *Neuron* 64 (4): 522–36.

LeDoux, J. (1994). "Emotional experience is an output of, not a cause of, emotional processing." In *The Nature of Emotion*, edited by P. Ekman and R. J. Davidson, 394–96. New York: Oxford University Press.

LeDoux, J. (1996). *The Emotional Brain*. New York: Simon and Schuster.

LeDoux, J. (2000). "Emotion circuits in the brain." *Annual Review of Neuroscience* 23: 155–84.

LeDoux, J. (2007). "The amygdala." *Current Biology* 17: R868–R874.

LeDoux, J. (2012). "Rethinking the emotional brain." *Neuron* 73 (4): 653–76.

LeDoux, J. (2015). *Anxious: Using the Brain to Understand and Treat Fear and Anxiety*. New York: Viking.

LeDoux, J. (2017). "Semantics, surplus meaning, and the science of fear." *Trends in Cognitive Sciences* 21: 303–6.

LeDoux, J., and R. Brown. (2017). "A higher-order theory of emotional consciousness." *PNAS*: E2016–E2025.

LeDoux, J., and A. R. Damasio. (2013). "Emotions and Feelings." In *Principles of Neural Science*, edited by E. R. Kandel, J. H. Schwartz, T. M. Jessell, S. A. Siegelbaum, and A. J. Hudspeth, 1079–94. New York: McGraw-Hill.

LeDoux, J., and D. S. Pine. (2016). "Using neuroscience to help understand fear and anxiety: A two-system framework." *American Journal of Psychiatry*: appiajp201616030353.

Liang, Z., G. D. R. Watson, K. D. Alloway, G. Lee, T. Neuberger, and N. Zhang. (2015). "Mapping the functional network of medial prefrontal cortex by combining optogenetics and fMRI in awake rats." *Neuroimage* 117: 114–23.

Lindquist, K. A., T. D. Wager, H. Kober, E. Bliss-Moreau, and L. Feldman Barrett. (2012). "The brain basis of emotion: A meta-analytic review." *Behavioral and Brain Sciences* 35: 121–43.

Lufityanto, G., C. Donkin, and J. Pearson. (2016). "Measuring intuition: Nonconscious emotional information boosts decision accuracy and confidence." *Psychological Science* 27: 622–34.

Lynn, S. K., and L. Feldman Barrett. (2014). "'Utilizing' signal detection theory." *Psychological Science* 25.

MacMillan, M. (2000). *An Odd Kind of Fame: Stories of Phineas Gage*. Cambridge, MA: MIT Press.

Mantini, D., U. Hasson, V. Betti, M. G. Perrucci, G. L. Romani, M. Corbetta, G. Orban, and W. Vanduffel. (2012). "Interspecies activity correlations reveal functional correspondence between monkey and human brain areas." *Nature Methods* 9: 277–82.

Marr, D. (1982). *Vision: A Computational Investigation into the Human Representation and Processing of Visual Information*. New York: W. H. Freeman.

Mayberg, H. S., A. M. Lozano, V. Voon, H. E. McNeely, D. Seminowicz, C. Hamani, J. M. Schwalb, and S. H. Kennedy. (2005). "Deep brain stimulation for treatment-resistant depression." *Neuron* 45: 651–60.

Medina, J. F., C. Repa, M. D. Mauk, and J. E. LeDoux. (2002). "Parallels between cerebellum- and amygdala-dependent conditioning." *Nature Reviews Neuroscience* 3: 122–31.

Mendl, M., O. H. P. Burman, and E. S. Paul. (2010). "An integrative and functional framework for the study of animal emotion and mood." *Proceedings of the Royal Society B: Biological Sciences* 277: 2895–904.

Millikan, R. G. (1989). "In defence of proper functions." *Philosophy of Science* 56: 288–302.

Mobbs, D., C. Hagan, T. Dalgleish, B. Stilson, and C. Prevost. (2015). "The ecology of human fear: Survival optimization and the nervous system." *Frontiers in Neuroscience: Evolutionary Psychology and Neuroscience* 9: 55.

Mobbs, D., P. Petrovic, J. L. Marchant, D. Hassabis, N. Weiskopf, B. Seymour, R. Dolan, and C. Frith. (2007). "When fear is near: Threat imminence elicits prefrontal-periaqueductal gray shifts in humans." *Science* 317: 1079–83.

Mobbs, D., R. Yu, J. B. Rowe, H. Eich, O. Feldman Hall, and T. Dalgleish. (2010). "Neural activity associated with monitoring the oscillating threat value of a tarantula." *PNAS* 107: 20582–86.

Morrison, S. E., A. Saez, B. Lau, and C. D. Salzman. (2011). "Different time courses for learning-related changes in amygdala and orbitofrontal cortex." *Neuron* 71: 1127–40.

Motzkin, J. C., C. L. Philippi, R. C. Wolf, M. K. Baskaya, and M. Koenigs. (2015). "Ventromedial prefrontal cortex is critical for the regulation of amygdala activity in humans." *Biological Psychiatry* 77: 276–84.

Müri, R. M. (2015). "Cortical control of facial expression." *Journal of Comparative Neurology* 524: 1578–85.

Namburi, P., A. Beyeler, S. Yorozu, G. G. Calhoon, S. A. Halbert, R. Wichmann, S. S. Holden, K. L. Mertens, M. Anahtar, A. C. Felix-Ortiz, et al. (2015). "A circuit mechanism for differentiating positive and negative associations." *Nature* 520: 675–78.

Nesse, R. M., and E. Ellsworth. (2009). "Evolution, emotions, and emotional disorders." *American Psychologist* 64: 129–39.

Niewenhuys, R. (2012). "The insular cortex: A review." *Progress in Brain Research* 195: 123–63.

Noe, A. (2004). *Action in Perception*. Cambridge, MA: MIT Press.

Nummenmaa, L., E. Glerean, R. Hari, and J. K. Hietanen. (2014). "Bodily maps of emotions." *PNAS* 111: 646–51.

Nummenmaa, L., and H. Saarimäki. (2017, in press). "Emotions as discrete patterns of systemic activity." *Neuroscience Letters*.

Ochsner, K. N., and J. J. Gross. (2005). "The cognitive control of emotions." *TICS* 9: 242–49.

Owald, D., and S. Waddell. (2015). "Olfactory learning skews mushroom body output pathways to steer behavioral choice in *Drosophila*." *Current Opinion in Neurobiology* 35: 178–84.

Owen, A. M., M. R. Coleman, M. Boly, M. H. Davis, S. Laureys, and J. D. Pickard. (2006). "Detecting awareness in the vegetative state." *Science* 313: 1402.

Oya, H., M. A. Howard, V. A. Magnotta, A. Kruger, T. D. Griffiths, L. Lemieux, D. W. Carmichael, C. I. Petkov, H. Kawasaki, C. K. Kovach, et al. (2017). "Mapping effective connectivity in the human brain with concurrent intracranial electrical stimulation and BOLD-fMRI." *Journal of Neuroscience Methods* 277: 101–12.

Panksepp, J. (1998). *Affective Neuroscience*. New York: Oxford University Press.

Panksepp, J. (2011a). "The basic emotional circuits of mammalian brains: Do animals have affective lives?" *Neuroscience and Biobehavioral Reviews* 35: 1791–804.

Panksepp, J. (2011b). "Cross-species affective neuroscience decoding of the primal affective experiences of humans and related animals." *PLoS ONE* 6: e21236.

Parr, L. A., and B. M. Waller. (2006). "Understanding chimpanzee facial expression: Insights into the evolution of communication." *Social Cognitive and Affective Neuroscience* 1: 221–28.

Parvizi, J., S. W. Anderson, C. O. Martin, H. Damasio, and A. Damasio. (2001). "Pathological laughter and crying: A link to the cerebellum." *Brain* 124: 1708–19.

Paton, J. J., M. A. Belova, S. E. Morrison, and C. D. Salzman. (2006). "The primate amygdala represents the positive and negative value of visual stimuli during learning." *Nature* 439: 865–70.

Pessoa, L. (2013). *The Cognitive-Emotional Brain: From Interactions to Integration*. Cambridge, MA: MIT Press.

Pfaff, D., L. Westberg, and L. Kow. (2005). "Generalized arousal of mammalian central nervous system." *Journal of Comparative Neurology* 493 (1): 86–91.

Poldrack, R. (2006). "Can cognitive processes be inferred from neuroimaging data." *Trends in Cognitive Sciences* 10: 59–63.

Poldrack, R. A. (2011). "Inferring mental states from neuroimaging data: From reverse inference to large-scale decoding." *Neuron* 72: 692–97.

Pollack, S. D., and D. J. Kistler. (2002). "Early experience is associated with the development of categorical representations for facial expressions of emotion." *PNAS* 99: 9072–76.

Prinz, J. (2004). *Gut Reactions*. New York: Oxford University Press.

Quirk, G. J., and D. Mueller. (2008). "Neural mechanisms of extinction learning and retrieval." *Neuropsychopharmacology* 33: 56–72.

Ratcliff, R., and G. McKoon. (2008). "The diffusion decision model: Theory and data for two-choice decision tasks." *Neural Computation* 20: 873–922.

Reyn, C. R. von, P. Breads, M. Y. Peek, G. Z. Zheng, W. R Williamson, A. L. Yee, A. Leonardo, and G. M. Card. (2014). "A spike-timing mechanism for action selection." *Nature Neuroscience* 17: 962–70.

Rolls, E. T. (1999). *The Brain and Emotion*. New York: Oxford University Press.

Rosen, J. B. (2004). "The neurobiology of conditioned and unconditioned fear: A neurobehavioral system analysis of the amygdala." *Behavioral and Cognitive Neuroscience Reviews* 3: 23–41.

Rozin, P. (1996). "Towards a psychology of food and eating: From motivation to module to marker, morality, meaning, and metaphor." *Current Directions in Psychological Science* 5: 18–24.

Russell, J. A. (1980). "A circumplex model of affect." *Journal of Personality and Social Psychology* 39: 1161–78.

Russell, J. A. (2003). "Core affect and the psychological construction of emotion." *Psychological Review* 110: 145–72.

Rychlowska, M., R. E. Jack, O. G. B. Garrod, P. G. Schyns, J. D. Martin, and P. M. Niedenthal. (2017). "Functional smiles: Tools for love, sympathy, and war." *Psychological Science* 28 (9): 1259–70.

Safina, C. (2016). *Beyond Words: What Animals Think and Feel*. New York: Picador.

Salzman, C. D., and S. Fusi. (2010). "Emotion, cognition, and mental state representation in amygdala and prefrontal cortex." *Annual Review of Neuroscience* 33: 173–202.

Sapolsky, R. (2009). "Any kind of mother in a storm." *Nature Neuroscience* 12: 1355–56.

Scarantino, A. (2014). "The motivational theory of emotions." In *Moral Psychology and Human Agency*, edited by D. Jacobson and J. D'Arms. New York: Oxford University Press.

Scarantino, A. (2016). "The philosophy of emotions." In *Handbook of Emotions*, 4th ed., edited by L. Feldman Barrett, M. Lewis, and J. M. Haviland-Jones. New York: Guilford Press.

Scherer, K. R. (1984). "On the nature and function of emotion: A component process approach." In *Approaches to Emotion*, edited by K. R. Scherer and P. Ekman. Hillsdale, NJ: L. Erlbaum Associates.

Scherer, K. R. (1988). "Criteria for emotion-antecedent appraisal: A review." In *Cognitive Perspectives on Emotion and Motivation*, edited by V. Hamilton, G. H. Bower, and N. H. Frijda, 89–126. Dordrecht: Nijhoff.

Scherer, K. R. (1994). "Emotions serve to decouple stimulus and response." In *The Nature of Emotion*, edited by P. Ekman and R. J. Davidson, 127–30. New York: Oxford University Press.

Scherer, K. R. (2009). "The dynamic architecture of emotion: Evidence for the component process model." *Cognition and Emotion* 23: 1307–51.

Schirmer, A., and R. Adolphs. (2017). "Emotion perception from face, voice, and touch: Comparisons and convergence." *Trends in Cognitive Sciences* 21: 216–28.

Searle, J. (1980). "Minds, brains, and programs." *Behavioral and Brain Sciences* 3: 417–57.

Seth, A. K. (2013). "Interoceptive inference, emotion, and the embodied self." *Trends in Cognitive Science* 17: 565–73.

Shackman, A. J., and A. S. Fox. (2016). "Contributions of the central extended amygdala to fear and anxiety." *Journal of Neuroscience* 36: 8050–63.

Shackman, A. J., A. S. Fox, R. C. Lapate, and R. J. Davidson, eds. (2018). *The Nature of Emotion: Fundamental Questions*. 2nd ed. New York: Oxford University Press.

Sharot, T. (2011). *The Optimism Bias*. New York: Vintage Books.

Shiffrin, R. M., and W. Schneider. (1977). "Controlled and automatic human information processing: II. Perceptual learning, automatic attending, and a general theory." *Psychological Review* 84: 127–91.

Siegel, E. H., M. K Sands, W. Van den Noortgate, P. Condon, Y. Chang, J. Dy, K. S. Quigley, and L. Feldman Barrett. (In press). "Emotion fingerprints or emotion populations? A meta-analytic investigation of autonomic nervous system features of emotion categories." *Psychological Bulletin.*

Simon, H. A. (1967). "Motivational and emotional controls of cognition." *Psychological Review* 74: 29–39.

Skerry, A. E., and R. Saxe. (2015). "Neural representations of emotion are organized around abstract event features." *Current Biology* 25: 1–10.

Smith, T. W. (2016). *The Book of Human Emotions: An Encyclopedia of Feeling from Anger to Wanderlust.* London: Wellcome Collection.

Spunt, R., and R. Adolphs. (2017a, in press). "The neuroscience of understanding the emotions of others." *Neuroscience Letters.*

Spunt, R., and R. Adolphs. (2017b). "A new look at domain-specificity: Lessons from social neuroscience." *Nature Reviews Neuroscience* 18: 559–67.

Sternson, S. M., and A.-K. Eiselt. (2017). "Three pillars for the neural control of appetite." *Annual Review of Physiology* 79: 401–23.

Sullivan, R., and D. Wilson. (2017). "Neurobiology of infant attachment." *Annual Review of Psychology* 69.

Susskind, J. M., D. H. Lee, A. Cusi, R. Feiman, W. Grabski, and A. K. Anderson. (2008). "Expressing fear enhances sensory acquisition." *Nature Neuroscience* 11: 843–50.

Swanson, L. W. (2005). "Anatomy of the soul as reflected in the cerebral hemispheres: Neural circuits underlying voluntary control of basic motivated behaviors." *Journal of Comparative Neurology* 493: 122–31.

Tarr, M. J., and I. Gauthier. (2000). "FFA: A flexible fusiform area for subordinate-level visual processing automatized by expertise." *Nature Neuroscience* 3: 764–69.

Tinbergen, N. (1951). *The Study of Instinct.* Oxford: Clarendon Press.

Todorov, A. (2017). *Face Value: The Irresistible Influence of First Impressions.* Princeton, NJ: Princeton University Press.

Tovote, P., J. P. Fadok, and A. Lüthi. (2015). "Neuronal circuits for fear and anxiety." *Nature Reviews Neuroscience* 16: 317–31.

Tybur, J. M., D. Lieberman, R. Kurzban, and P. DeScioli. (2013). "Disgust: Evolved function and structure." *Psychological Review* 120: 65–84.

Wang, L., I. Z. Chen, and D. Lin. (2015). "Collateral pathways from the ventromedial hypothalamus mediate defensive behaviors." *Neuron* 85: 1344–58.

Winkielman, P., and K. C. Berridge. (2004). "Unconscious emotions." *Current Directions in Psychological Science* 13: 120–23.

Woo, C. W., L. Schmidt, A. Krishnan, M. Jepma, M. Roy, M. A. Lindquist, L. Y. Atlas, and T. D. Wager. (2017). "Quantifying cerebral contributions to pain beyond nociception." *Nature Communications:* doi: 10.1038/ncomms14211.

Yarkoni, T., R. Poldrack, R. Nichols, D. C. Van Essen, and T. D. Wager. (2011). "Large-scale automated synthesis of human functional neuroimaging data." *Nature Methods* 8: 665–70.

Yilmaz, M., and M. Meister. (2013). "Rapid innate defensive responses of mice to looming visual stimuli." *Current Biology* 23: 2011–15.

Zadbood, A., J., Y. C. Chen, K. Leong, K. A. Norman, and U. Hasson. (2017). "How we transmit memories to other brains: Constructing shared neural representations via communication." *Cerebral Cortex:* doi: 10.1093/cercor/bhx202.

Zeng, H., and J. R. Sanes, (2017). "Neuronal cell-type classification: Challenges, opportunities, and the path forward." *Nature Reviews Neuroscience* 18: 530–46.

INDEX

Note: Illustrations are indicated with **bold** page numbers.

abstraction, levels of, 115–18, **117**, 126, 143

age or developmental stage: and emotions, 20–21, 167–68; and fear conditioning studies, 165; and regulation of emotion, 89

aggression and anger, 209, 220–21, **275**, 300; expression of, **91**; and hypothalamus, 182–83, 188–89, 194, 196; in insects, 209–10; and mating behaviors, 209–10; Panksepp's RAGE system, 193–94, 300–301; in rodents, 161–62, 183, 193–94

alexithymia, 305–6

algorithmic models, 118–19, 148; and machine learning, 239–40

amnesia, 75–76

amygdala: defensive behaviors and, 164, 177, 188, 195, 196, 247; distributed neuronal activity in, 176; and emotion, 231, **232**; and fear, 10–11, 32–33, 37, 163–64, **318**; fear conditioning and the, 169–73, **170**, **173**, 181, 195; and global coordination, 42, 171, 181–82, **318**; interactions with other brain structures, 42, 163–64; LeDoux on fear and the, 163–64; mouse vs. human amygdala activation, **175**; neuronal subpopulations in, 171–73, **173**, 179, 195; nuclei involved in fear conditioning, 169–71, **170**; olfactory processing in, 169; open questions for future research, 317–20; prefrontal cortex and regulation of, 178–79, 198; and rapid processing of cues, 258; research biases and focus on, 231, **232**; role in rodent fear responses, **34**, 164–179, **175**; social and reproductive behaviors and the, 169; and valence of responses, 174–76. See also basolateral amygdala (BLA)

anger. See aggression and anger

animals, emotional behavior in: and building blocks of emotion, 62; and dogs as subjects, 247–50, **249**; and emotions as human subjective experience, 127, 159; and evolution of emotion (see building blocks); expression of emotion in, **91**; π states, 144; social communication and, 90–93. See also specific animals, e.g., rodents; specific behaviors, e.g., anxiety

Animal Sentience (journal), 311

animal studies: and age or maturation of subjects, 165; and distinction between feeling and emotions, **145**, 153; and emotion as uniquely human phenomena, 127; genetic model organisms and, 114, 126; and genetic or species variation, 180; and homologies across species, 124–25, 154–55, **170**, 172, **175**, 187, 197–98, 247, 248, 264, 277–78, 294–95, 300–301, 314, 315, 322–23; and imaging technologies, 132, 136–38, 176; and individual differences, 322–23, **323**; invasive experimental methods and, 131–32; nonhuman primates (NHPs) as subjects, 156–57; and psychiatric drug research and development, 153–55; and research into psychiatric disorders, 134; and species-specific emotions, 321–22; transgenic (genetically engineered) subjects, 138–40, 156–57; value of, 128, 130, 131–34, 212, 214, 277–78, 310. See also specific, e.g., rodents

anterior cingulate cortex, **224**, 271, **271**, 287

anxiety, 48; as anticipatory rather than reactive, 189–90; anxiety disorders, 189–90; circuits for, **192**; as distinct from fear, 190–91, **192**; drugs and treatment of, 153–55, 190 (see also anxiolytic drugs); in insects and other arthropods, 208–9; as malfunction, 47–48, 190

anxiolytic drugs, 153–55, **155**, 160, 208–9, 324

appetitive behaviors, 173, **173**, **175**, 195, 196; aversive vs. appetitive conditioning, 174, 199–200, **202**; valence and, **66**

appraisal theory, 55, 87–88, 89, 260, 288–90; "appraisal 38," **261**, 262–63

Arnold, Magda, 288

arousal, 22–24, 127, 145–48; core affect and, 69–70, 99, 283–84, 290 (*see also* 2-D space for mapping emotions); as distinct from intensity, 66; and fear, 33; neuromodulators and, 146–48; persistence of, 147; role of hormones and neuropeptides in, 146–47; and scalability (intensity), 69, 147; threat alert or defensive, 206, 212–14; and valence, 69–71, 307. *See also* MAD states (motivation, arousal, and drive states)

associative learning, 77, 164–65, 169; consolidation of learned fear, 178; and domain specificity, 98; dopamine and, 200–201; extinction learning, 178–79; fear conditioning, 164–65, **166**, 168–69; and generalization, 77; in insects, 198–203, 211; learned responses to stimuli, 161–62; motivation and, 152; Pavlovian conditioning, 77, 98, 164–65, **166**, 195; social learning, 77

attachment in infants or animal offspring, 167–68

attribution of emotion to others, 96, 257–60, **259**, 264, 278, 299, 311

automaticity, 23, 62, 66, 99; and dual process-theories of cognition, 81–86; and social communication, 90–97; volitional control vs., 83–84

autonomic functions and emotions, 35

aversive conditioning, 163, 167, 174, 199–200, **202**, 214, 272–73

Bargmann, Cornelia, 211–12

basic emotions: as characters in *Inside Out,* 5–6; culture and universality of, 6; Ekman and, 6, 94, 256, 260, 301; as irreducible, 6–7; Panksepp and, 294

basolateral amygdala (BLA): cellular heterogeneity in, 174–76; fear conditioning and the, 169–70, **170**; imaging

technology and activity in, **175**; neuronal subpopulations in, **173**, 179, 195

bed nucleus of stria terminalis (BNST), 181, 183, **184**, 190–91, 195, 271–72

bees, 87, 210–11

behavior: behavioral assays in rodents, 154–55, 190–91, **192**; as cause or consequence of emotion, 9–10, 24; emotions and control of, 9–10, 38–39; as expression of emotion, 4–5, 9–10, **41**, 42, 60, 92–93, **145**; genes and control of, 198–99; global coordination and, 78–79; as goal-oriented, 72, 148, 151, 167–68, 296. *See also specific, e.g.,* defensive behaviors

behaviorism, 9–10

Bekoff, Marc, 311–13

Benzer, Seymour, 198–99

Berns, Greg, 248

Berridge, Kent, 303, 306

Beyond Words: What Animals think and Feel (Safina), 311

biological organization in brain: contrasted with levels of abstraction, 116; scales of, 105–8, **107**, 126

"blindsight," 51

The Book of Human Emotions (Watt), 7

Boorstin, Ralph J., 1

The Brain and Emotion (Rolls), 70

brain chemistry, 31, 214; emotions and, 117–18. *See also* neuromodulators

brain regions: anatomical segregation and specific, 230; and attribution of emotion to others, 96, 257–60, **259**, 299; coordination and communication among, 37–38, 79, 80, 140–41, 170–71, 179, 223, 240–41, 276, 309, 317–19; and coordination of emotion, 37–38, 140–41, 317–19; and facial expressions, 93; and generation of emotion, 10–11, 37, 140–41; homologies across species, 124–25, 154–55, 172, **175**, 187, 197–98, 247, 248, 264, 294–95, 300–301, 314, 315, 322–23; and human emotion, **224**; "inner brain" and emotion, 113; invertebrate brain structures, 197–98; involved in different types of fear, **184**; and localization of emotion systems, 10–11, 142, 188–89, 223–26, 230–33, 249; number of distinct, 179; open questions for future research, 314–17; populations of neurons contrasted with, 123–24, 133, 134–35; psychological functions and, 121; and similarity analysis, 243; structures of invertebrate nervous systems, 207–8; and studies of human emotion, 222–24; threat imminence and activation of, **272**. *See also specific, e.g.,* amygdala

Braitenberg, Valentino, 324–25

Brenner, Sydney, 211

building blocks: in cephalopods, 212–14; as "emotion primitives," 129, 143–44, 193–94, 203–6, **204**, 212, 309; evolution of emotion and, 143; invertebrate brains and, 198; and properties of emotion states, 62–63

C. elegans. See worms

calcium imaging, 126, 136–40, 309

categories of emotion (taxonomy), 7–8, **26**, 282, 293–94, 300

catFISH (cellular compartment analysis of temporal activity by Fluorescent In Situ Hybridization), 183–84

causality: animal studies and, 140–43; vs. causation, 111–14, **112**; correlation vs. causation, 126; as deterministic rather than probabilistic, 113–14; dynamic causal modeling, 241; and fMRI studies, 240–42, 267–68; mechanisms and, 109–10, 131–32; as mediated or indirect, 113; necessity and, 111–13; and relationships among brain regions, 229; sufficiency and, 111, 113–14; testing causal relationships, 114–15

causal models, **82**, **241**

cell type specificity, **109**; genetics and, 123; identification of cell types, 123

cephalopods: emotion primitives in, 213–14

c-fos, 183–86

chemogenetics, 126, 156–59

cingulate cortex, 223, **224**, 271, **271**, 287

cognitive bias in bees, 210–11

cognitive processing: influence of emotion on, 23, 33; mood as influence on, 72–73

component process model, 289–90

computer: as metaphor for brain, 116–20

conditioning. *See* associative learning

connectomes: for *C. elegans,* 211; Human Connectome Project, 239

consciousness: animals and conscious experience of emotion, 312; brain structures and, 197–98; categories of, 282; and distinction between emotion states and emotions, 269–70; emotion and, 49–51, 303, 308; and emotion in humans, 232–33, 283; feeling and conscious experiences of emotions, 281; invertebrates and, 197; LeDoux on animal, 297–98; and LeDoux's higher-order theory of emotion, 297–98; nonconscious emotions, 303, 304–7

constructed emotion theory (Feldman), 290–94, **292**, 302

construct validity, 190, 217–18, 253, 304

convergent validity, 217–18

coordination. *See* global coordination

core affect, **68**, 69–70, 99, 283–84, 290, 307

core relational themes, 55, 289–90

correlation, 273–74; vs. causation, 111–14, **112**, 126

Craig, Bud, 286–87

crayfish, 209

cultural components of emotion, 7, 21, 264–65; and bodily, "felt" locations of emotion, 36; facial expressions and, 94, 95

Damasio, Antonio, 118–19, 219, 226–28, 268, 286–88, 298

Damasio, Hanna, 219

Darwin, Charles, 42, 53, 95, 143, 197–98, 206, 259–60, 301, 308, 309–10; and behavior as expression of emotion, **145**; and "goal oriented" behavior linked to emotion, 151

Davis, Michael, 190–91

decision making: cognitive bias and, 210–11; drift-diffusion models and, 73–74, 84–85, 99; dual process theories of, **84**–85; effects of emotion states on, 33, 38, 84–85, 86–87, 186–87, 190, 211, 303, 305; fight-or-flight as decision-making process, 38, 55, 60, 163, 164, 171; in flies, 206–7; persistence and, 73–74

deep brain stimulation (DBS), 31, 115, 136

defense conditioning. *See* fear conditioning
defensive arousal or threat alert, 206, 212
defensive behaviors: amygdala and, 164,
177, 188, 195, 196, 247; in cephalopods,
213–14; defensive conditioning, 165;
fear and, 53, 164–65, 187, 311; hypothalamus and, 142, 187, 196, 314; as innate,
162–63, 179–89, **204**; in insects, **204**,
206–8, 214; mechanisms for innate vs.
conditioned, 181–82; in rodents, 52–56,
56, 154, 179–83; ventromedial hypothalamus (VMH) and mediation of,
182–83. *See also* aggression and anger
directed graphs, 80–82
discriminative validity, 217–18
disgust: as "basic" emotion, 5–6, 94, 291;
bodily map of, **36**; as constructed emotion, 291; functional varieties of, **21**;
the insula and, 223–25, 229–30, 249;
as learned reaction, 21, 97–98; stimuli
and, 64, 97, 304
distributed neuronal or cellular activity, 37,
133; in amygdala, 176; and coordination
among brain regions, 179; and global
coordination of emotion states, 80; vs.
localized emotions in the brain, 189–90
dogs, fMRI studies and, 247–50, **249**
domain-general processes, 291, **292**
domain specificity, 63, 98
dopamine, neurotransmitter, 117, 146–47,
200–203, **202**
dorsomedial prefrontal cortex (dmPFC),
96, **259**, 259–60, 263, 276, 278
"double dissociation," 190–91
DREADDs, 135, 159
drift-diffusion models, 73–74, 84–85, 99
drive, 22–24, 143–44, **145**, 159; contrasted
with motivation, 148–50; as "hydraulic"
internal state, **149**, 149–50
Drosophila. See insects
dual process-theories of cognition, 81–84,
99
Duff, Melissa, 75

ecological validity, 179, 203, 251, 253, 256,
278
EEG (electroencephalography), 78, 233
Ekman, Paul, 6, 94
electrical stimulation studies, 169, 182,
229–30, 249, 315; and amygdala, 169;

deep brain stimulation (DBS), 31, 115,
136; in humans, 182, 187, 229–30
emergent properties, 106–7
emotion: attribution of emotion to
others, 96, 257–60, **258**, **259**, 264,
278, 299; behavior as expression of,
4–5, 9–10, **41**, 42, 60, 92–93, **145**; as
biological phenomena, 308; bodily
maps of where emotions are "felt," **36**;
as both innate and learned, 99; brain
chemistry and, 117–18; brains regions
or structures as "location" of, 10–11,
24–25; categories of (taxonomy), 7–8;
cognitive processing influenced by, 23;
common misconceptions of, 1, 4–13;
consciousness and, 49–51, 182, 232–33,
281, 283, 303–7, 308; as consequence
of behavior, 9–10, 24; contrasted with
motivation, 150–53; control of (*see*
emotion regulation or control); deliberative behavior as distinct from, 22;
as distinct states, 62; evolution of (*see*
evolution of emotion); "feelings" as
distinct from, 12, 24, 49–51, 56, **145**,
153–54; and flexibility of response, 64;
as internal brain states (*see* emotion
states); as "interrupt," 85; learning as
property of, 62, 64; as localized in regions of the brain, 10–11, 142, 188–89,
223–26, 230–33, 249; memory and,
22–23, 75–76, 276–77; nonconscious,
303–7; persistence of, 62; as processes
rather than fixed states, 18; reflex and,
8–9, 71; as response to stimuli, 8–9,
24, 161–62; social function of, 301–2;
as species specific, 321–22; as state
(*see* emotion states); as subjective
experience, 24; taxonomy of, **26**, 30,
282, 293–94, 300; as uniquely human
phenomena, 127
The Emotional Lives of Animals (Bekoff),
311
emotion primitives. *See* building blocks
emotion regulation or control: as age
dependent, 85; in animals, 86; cognitive processes and control of, 38–39;
as evolutionary adaptation, 89–90;
as metacognitive ability, 87–90, **88**;
and prefrontal cortex (PFC), 85, 89,
223; and reappraisal, 89; as volitional,
84–85, **88**

emotion states: central, 270–73; circuits mediating innate behaviors linked to, 182–88; dissociating concepts or experience from, 273–78; as distinct from experience of emotion, 308, 310; in *Drosophila,* 203–8; as functionally defined, 99; hypothalamus and coordination of, 188–89; mapping similarity among, 67–71; as π states, 144; relationship to MAD (Motivation, Arousal, and Drive) states, 143–53; spatiotemporal scope of, **59**; valence and, 203

endocrine responses, 33, **145**, **155**, 189–90

enhancers, gene expression and, 138

Epstein, Robert, 119

evolution of emotion, 7–8, 197–98, 281, 309–10; and adaptation to environmental challenge, 40–42, 47, 53–54, 59–60, 309–10; adaptive value of emotion-like states, 206–7; and co-option of pre-existing mechanisms, 93; and emotions as decoupled reflexes, 19–22; as flexible central states, 20–22; invertebrate brains and, 197–98; selection and innate representations of valence, 201–3; and social communication, 93. *See also* building blocks

expression of emotion: in animals, **91**, 128; behavior as, 4–5, 9–10, **41**, 42, 60, 92–93, **145**; inference and observation of, 128

The Expression of the Emotions in Man and Animals (Darwin), 42, 259–60, 309

extinction learning, 178–79

Facial Affect Coding System (FACS), 96

facial expressions: amygdala and processing of, 258–59; and attribution of emotion to others, **228**, **258**, **259**; Ekman and expression of basic emotions via, 6, 94, 256, 260; visual processing of, 64–65, 93–97

Fanselow, Michael, 162–63

fear: adaptive and maladaptive, 47–49; amygdala and, 10–11, 32–33, 37, 169–73, **318**; and arousal, 33; brain regions involved in, **184**; defensive behaviors and, 53, 164–65, 179–89, 311; as distinct from anxiety, 190–91, **192**; expression, 168–71; "fight-or-flight" responses, 38,

55, 60, 163, 164, 171; as innate, 180–81; as learned response (*see* fear conditioning); lesions and dissociations of fear, 226–28; pathways mediating, 181–83, **184**; and threat processing, 52–55, **54**, 59–60; typology of fear states, 191–93

fear conditioning, 164, 168–71; and cellular heterogeneous cell populations in amygdala, 171–73, **173**; fear extinction, 178; in insects, 199–200; neuropeptides and, 177–78; and olfactory cues, 199–200; and persistence or consolidation, 177–78; in rodents, 164–79, 195

feelings: and animals, 129–30, 153–54; bodily maps of where emotions are "felt," **36**; as distinct from emotions, 12, 24, 49–51, 56, **145**, 153–54, 308; interoceptive theories of, 286–88; psychological theories about emotion and, 281; structure of affect and, 283–86; valence of daily experiences, **285**

Feinstein, Justin, 75

Feldman Barrett, Lisa, 6–7, 188–89, 290, 302, 307

Feynman, Richard, 324, **326**

"fight-or-flight" responses, 38, 55, 60, 163, 164, 171

flexibility of emotions, 99; and learning, 201–3

flies. *See* insects

fMRI (functional MRI): in animals, 245–50; and causality, 240–42; event-related studies, 252–53; faces as stimuli, 256–57; false positive findings and, 235, 237; limitations of, 234–35, 238, 245, 246, 251, 254, 278; meta-analysis, 231, **232**, 235–37; methodology, 233–41; and multivariate analysis, 238–39, 260–63, **261**; music as stimuli in studies, 253–56, 266–67; and network-level connectivity, 238; non-human primates and, 246; and region-of-interest (ROI) focus, 237–38; and similarity space analyses, 242–45; stimuli selection for studies, 253–57, 266, 270–73, 278

Fodor, Jerry, 63

"freezing" behavior: comparison of innate *versus* learned, 161–63; as distinct from periods of immobility, 180; and fear conditioning, 172

Fridja, Nico, 300

fruit flies. *See* insects
functional definition of emotion: causal relations and, 40–41, 56; consciousness and, 49–50; environmental challenges and, 52–53; functional properties (*see* properties of emotion states); and mechanisms of implementation, 45–46; and problem of "type identity," 44–45; and psychiatric disorders as "malfunction," 47–48, 56, 190; psycho-functionalism and, 40–41
functional level (computational level) of abstraction, 116
functional MRI. *See* fMRI (functional MRI)

Gage, Phineas, 219–20
gain-of-function and sufficiency, 113
GECIs (genetically encoded calcium indicators), 137–38
generalization, 66, 76–78
genetically encoded neuronal indicator or effector (GENIE), 138–40
genetic modification: and ablation of neuronal populations, 183, 189; gene therapies and, 157–58; genome editing technology, 157, 177–78; optogenetics, 113, 114–15, 123, 134–36, 156–59, 176, 323; and transgenic animal models, 138–40, 156–57; using viral vectors, 157–59, 316
GENIEs (genetically encoded neuronal indicators or effectors), 138–40
GEVIs (genetically encoded voltage indicators), 137–38
global coordination (global organismal response), 78–81, 99, 165; amygdala and, 171; fear and, 78–79, 171; and hierarchical vs. distributed control, 80; hypothalamus and, 187–89; mechanisms or architectures for, 79–80; open questions for future research, 317–20
goal directed, emotional behavior as, 72, 148, 151, 167–68, 296
Goldman, Alvin, 299
G-protein coupled receptors (GPCRs), 177–78
Griffiths, Paul, 300–302
Gross, James, 86, 90

Haidt, Jonathan, 21
Hall, Jeff, 199
Hanlon, Roger, 213
Hasson, Uri, 246–47
Heberlein, Ulrike, 209
Heisenberg, Martin, 200
Hess, Walter, 142, 194
higher-order theory of emotion (LeDoux), 297–99
homeostatic states, 149–51, 188, 287, 296; distinguished from emotion states, 151; hypothalamus and control over, 188
homologous recombination, 157
homunculus fallacy, 11–13
Human Connectome Project, 239
human neuroscience: brain regions of focus in, **224**; and conceptualization of emotions, 260–65, **261**, 283; consciousness and emotion in, 232–33; and construct validity, 217–18; electrical stimulation studies, 182, 187, 229–33, 249; lesion studies, 10, 115, 178–79, 218–28, 230–31, 234, 249, 287, 317–20; and metacognition, 283; methods and tools available to study, 215–17; and music as stimulus, 253–57, 266–67; regulatory barriers and, 157–58; and self-reported subjective experiences, 214, 216–17, 266, 310–11, 313
hypothalamus, 33, **224**; and aggression, 194, 196; and coordination of emotion states, 187–89, 317; and defensive states, 142, 182–83, 187–88, 196; and homeostatic functions, 187–88; and innate behavior, 196; role in orchestrated fear response, **34**; ventromedial hypothalamus (VMH), 163, 182–83, **184**, 196

immediate early genes (IEGs), 183–86
implementation level of abstraction, 116
individual differences, 263, **323**; fMRI studies and, 246–47; genetic basis of, 47; and unique emotions, 322–24
induction of emotion, 31–32, 38
inference, 128–30
inhibitory neurons, 132, 138–39, 178–79, 185
innate response to stimuli, 161
inputs. *See* stimuli

insects: associative learning in, 198–201, 198–203, 210, 211, 214; bees and cognitive bias, 210–11; brains contrasted with mammalian brains, 207; emotion states in, 203–8; mushroom bodies and MBONs, 201, **202**; nervous system in invertebrates, 200–201; persistence and emotion-like states in, 71–72

Inside Out (film), 4–13

insula, 120, 223–25, **224**, 229–31, 233–34, 249, 254–55, 259, **259**, 269, 276, 278, 286–87

intercalated cell mass (ic or ITC), 178–79

interdisciplinary approach, 16–18, **17**, 22–23, 27, 32, 103–5

interoceptive signals, 149, 223

invertebrates: cephalopods, 212–14; and emotion primitives, 212; emotion states in, 210–13; and genetic modifications for study, 115, 126, 134; jellyfish, 143, 198; as model organism, 134; nervous system contrasted with vertebrate, 200–201, 212; and optogenetics, 134; snails, 198, 199. *See also* insects; worms

irreducibility of emotions, 6–7

James, William, 9, 35, 120, 255, 260, 286
jellyfish, 143, 198

Kenyon cells, 201, **202**
Konopka, Ron, 199
Lazarus, Richard, 55, 288
LeDoux, Joseph, 29–30, 50, 127, **145**; on amygdala and fear, 163–64; on animal studies and development of psychiatric drugs, 153–54; and "defensive conditioning," 164–65; and higher-order theory of emotion, 297–99

lesion studies, 10, **109**, 115, 178–79, 218–28, 230–31, 234, 249, 287, 317–20

levels of description, 126; abstraction and, 115–18; biological organization and, 105–8; of biological organization of the brain, **107**; and methods of investigating the brain, **109**; and study of emotion, 105–8

Lorenz, Konrad, 149–50

loss-of-function and sufficiency, 113

Lüthi, Andreas, 176

MAD states (motivation, arousal, and drive states), 143–48; emotions contrasted with, 143–**45**, 159; in insects, 209–10

mapping immediate early genes (IEGs), 185

Marr, David, 116

mating behaviors, 182–83, 187, 194–95; and aggression, 209–10; amygdala and, 169; and MAD states, 209

medial hypothalamic defensive system, 183

medial prefrontal cortex (mPFC): in humans, 219–33; in rodents, 178–79

MEG (magnetoencephalography), **109**, 233, 289–90

Meister, Markus, 52

memory: amnesia, 75–76; and constructed emotion theory, 290; emotion and, 23, 75–76, 276–77; and higher-order theory of emotion, 298–99; and learning in flies, 199, 210; and persistence of emotion states, 75–76; as stimulus, 268–69

meta-analysis: Neurosynth program and, 231, **232**; and quality of data, 236–37, 277

mice. *See* rodents

midbrain periaqueductal gray (PAG), 33, **34, 170**, 171, 182, **192, 224, 272**, 273, 278, **318**

Millikan, Ruth, 46–47, 302

mind, theory of, 96–97, 104–5, 257–58, 264. *See also* attribution of emotion to others

Mobbs, Dean, 55, 270

modularity, 63–64

The Modularity of Mind (Fodor), 63

monkeys. *See* nonhuman primates

moods, 72–73, 87

motivation: animal models and study of, 127; dopamine and reward, 200–201; dopamine and reward or punishment, 201; drive contrasted with, 148–50; emotion contrasted with, 150–53; properties of motivational states, 151–52; as state, 62, 127; and stimulus response relationships, 149–50

MRI (Magnetic Resonance Imaging), 215; animal studies and, 248–49, **249**; lesion studies and, 220–21, **228**; of mother and child, 13–15, **14**. *See also* fMRI (functional MRI)

mushroom bodies: as analogous to the hippocampus, 207; mushroom body output neurons (MBONs), 201, 202
music, as stimulus, 253–57, 266–67

narcolepsy, 147
necessity and sufficiency: causality and, 111–13, 123–26, 140, 159
nematodes. *See* worms
neuromodulators, 31, 117, 146, 185–86; and arousal, 146–48; and persistence, 212
neurons, **107**; ablation of neuronal populations, 183, 189; activation in amygdala, 37, 176; cell types and populations, 123–24; and distributed nature of emotions in the brain, 37, 80, 133, 176, 179, 189–90; genetically defined neuronal cell types, 138–40; GENIEs and, 138–40; inhibitory, 132, 138–39, 178–79, 185; neuronal specificity, 135–36; optogenetics and activation of populations of, 114–15, 135; Sf1+, 183, 186–89; subpopulations in amygdala, 171–73, **173**, 179, 195; subpopulations in rodents, 132–33, 171–72, 201, 308–9; testing single-neuron activation, 115
neuropeptides: fear conditioning and, 177–78
Neurosynth, 231, **232**
nonhuman primates: emotions in monkeys, 247; fMRI studies and, 246; optogenetics and, 157; transgenic, 157
normalcy, 123–25
Nummenmaa, Lauri, 36, 274, **275**

Occam's razor, 153–54, 160
octopamine, 200
octopuses, 212–14
olfactory cues: and fear conditioning, 199–200; Kenyon cells and, 201, **202**; as stimuli, 97, 181, 198–200
OpenfMRI project, 237
open question for future research, 313–26
optical imaging, 136–38
optogenetics, 114–15, 123, 134–36, 159, 309; in humans, 157–59; and neuronal specificity, 135; and nonhuman primate (monkey) studies, 156–57

orbitofrontal cortex, **224**, 254, 296
Owen, Adrian, 130

pain, 275, 286; animals and, 266; fMRI studies of perception of, 275–76; as homeostatic emotion, 150–51; independent pain signature, 276
Panksepp, Jaak, 6; and basic emotions in animals, 6, 193–94; emotion systems and neurobiological systems theory, 294–96, 303; and subjective feelings in animals, 129–30, 164
parsimony, principle of (Occam's razor), 153–54, 160
Patient S. M., 226–28
Pavlovian conditioning, 77, 98, 195; and "emotional" learning, 164–65; fear conditioning, **166**
periaqueductal gray (PAG), 33, **34**, **170**, 171, 182, **192**, **224**, **272**, 273, 278, **318**
persistence, 71–78; and arousal, 147; memory and, 75–76; as property of motivational states, 151–52; as scalar quantity similar to decay constant, 75
personality traits, 73
PET (positron emission tomography), **109**, 233, 247, 254, 268–69, 277–78
pharmacogenetics: animal models and research, 132, 153–**55**; genetic editing and drug research, 178; "silencing" and, 134–36, 183, 189
philosophy of emotion, 299–303, 307
Pine, Danny, 153–54
π states, 144
Poldrack, Russ, 237
Post-Traumatic Stress Disorder (PTSD), 47, 165, 178
predictability, 24–25, 31, 266–67
prefrontal cortex (PFC): and amygdala, 317–19; amygdala regulation by, 178–79, 198; and attribution of emotion to others, 96, 278; and control of emotion, 85, 89, 223; development in primates, 223; and emotion, **224**, 231, **232**; emotional impairment and damage to, 219–22; fMRI studies of, 223; and threat imminence, **272**, 273
"primary" emotions. *See* basic emotions
Prinz, Jesse, 299

properties of emotion states: animals and expression of, 90–93, 129; automaticity, 23, 62, 66, 81–85, 90–97, 99; and difference between building blocks and features, 62–63; generalization, 66, 76–78; generic list of, 65–66; global coordination and, 66, 78–81, 99; learning, 62, 64, 77, 97–99; open questions for future research, 320–21; persistence, 62, 69–71, 186; scalability (intensity), 66, 69, 71, 186; Sf1+ neurons activation and, 186–89; social communication, 66; valence, 62–63, 66, 69–71, 186

psychiatric disorders, 134; pharmacogenetics and, 135; post traumatic stress disorder (PTSD), 47–49, 56, 165

psycho-functionalism, 40–41

psychophysiology, 16–17; as diagnostic of emotion, 35

RAGE (Panksepp), 193–94, 300–301

Ratcliff, Roger, 73

rats. See rodents

reflexes: and domain specificity, 63–64; emotions as decoupled, 18–22; emotions as distinct from, 143; as involuntary behavior, 18–19, 20; reflexive jumping in insects, 208

representational similarity analysis (RSA), 242–45, **244**, **261**, 262–63, 266

responses to stimuli: comparison of innate *versus* learned, 161–63; as learned behaviors, 161

reward: and amygdala, 174–75; brain regions linked to, 195, 254; and dopamine, 200–201; Roll's theory, 296–97

robots, 143, 324–26; emotional AI, 24–25

rodents: aggressive behavior in, 161–62, 183, 193–94; and anxiety, 190–93; and attribution of emotion to others, 264; behavioral assays in, 154–55, 190–91, **192**; brain regions involved in fear, **184**; c-fos and IEGs in, 183–86; contrasted with non-human primates as study organisms, 156; defensive behaviors in, 52–56, **56**, 179–83; and drug research, 153–55; fear conditioning in, 164–79, 195; and fear extinction, 178–79; fMRI and, 246, **323**; and homologies with humans, 124–25, 154–55, 172, **175**, 187, 247, 264,

300–301, 314, 322–23; and individual differences, 253, 322–23, **323**; innate defensive behaviors in, 179–83; learned vs. innate behaviors in, 162–63; mouse vs. human amygdala activation, **175**; neuronal subpopulations in, 132–33, 171–72, 201, 308–9; number of anatomically distinct brain regions in, 179; optogenetics, 113, 134, 176, **323**; and orchestration of fear response, 34; RAGE (basic arousal state) in, 193, 300; role of amygdala in fear response of, 34, 164–179, **175**; and social communication of emotion, 262; threat assessment in, 52–56, **56**

Rolls, Edmund, 70, 296–97, 304, 307

Rosbash, Michael, 199

Rozin, Paul, 21

Ruta, Vanessa, 201

Safina, Carl, 311

Saxe, Rebecca, 13–15, 260–62

scalability, as property, 66, 71, 186; and arousal, 69, 147; of motivational states, 151–52

Scarantino, Andrea, 299–300

Scherer, Klaus, 55, 288–89, 307

Schneider, W., 83

Schnitzer, Mark, 176

Searle, John, 119

sensory cues: exteroceptive and interoceptive, 151; olfactory, 199–200; as unconditioned stimuli, 97

Seth, Anil, 291

sexual behaviors. See mating behaviors

Sf1+ neurons, 183, 186–89

Shiffrin, R. M., 83

silencing, pharmacogenetic, 158–59, 183, 189

similarity spaces, 67–69, 242–43, 274, **275**, 323–24; and imaging of emotion concepts, 260–63

Simon, Herbert, 85

Smith, Tiffany Watt, 7

snails, 198

social behaviors: amygdala and, 169; and attribution of emotions to others, 96; communication (*see* social communication); mating behaviors, 182–83, 187, 194–95, 209–10; VMH and, 182–83, **184**, 194 social communication: animals and, 90–93; and automaticity, 90–97; facial expressions and, 93–97; as a feature, 52–53; interspecies, 248–49

Spemann, Hans, 152

squid, 212–14

stimuli: domain specificity of, 63–65; ecologically valid, 179, 203, 251, 253, 256, 278; electric shocks as, 64, 167–68, 199, 272–73; emotion and response to, **41**, 42; emotions as decoupled reflexes, 18–22; humans and thoughts as, 283–84; inducing strong emotion states with realistic, 270–73, **271**; memories as, 268–69; modality of sensory input, 180–81; olfactory cues as, 181, 199–200; subliminal, 304–5

sufficiency: causality and, 111, 113–14; in psychological theories, 284

survival circuits, 297–98

synapses, **107**, **109**, 110, 200–201, 234

systems neurobiology, 104, 126

terms, use of, 221, 281; cognitive atlas as reference, 122; debates over use of word "emotion," 127–28; "defensive conditioning" vs. fear conditioning, 165; and distinction between emotions and feeling, 50–51; evolving understanding and, 152–53; interdisciplinarity and, 103–5, 120–23, 125

threat assessment and threat processing, **54**, 59–60; accumulation of sensory input and, 73–74; in cephalopods, 213; as cumulative, 186–87, **204**, 205–6; imminence and, 270–73, **271**; innate defensive responses in flies, 203–6, 214; in insects, 206–8, 214. *See also* defensive behaviors

Tinbergen, Niko, 154, 195

Tonegawa, Susumu, 173

Tranel, Daniel, 75, 226

transcranial magnetic stimulation (TMS), 115, 229–30, 233

transgenic animal models, 138–40, 156–57

Tsao, Doris, 246

2-D mapping schemes: core affect and emotion mapping, **68**, 69–70, 307; Roll's theory and, 297; shortcomings of, 70–71; similarity analyses and, 242

units of explanation, 110

Urbach-Wiethe disease, 226–28

valence: and conditioning of insects, 199–200; core affect and, **68**, 69–70, 99, 283–84, 290, 307; and decision making, 87; dysfunction and, 48–49; of everday experiences, **285**; and innate or conditioned responses, 201–3; positively valenced emotion states, 195; as property of emotion states, 62–63, 66, 69–71, 186; as property of motivational states, 151–52; psychological use of term, 284; and responses in amygdala, 174–76

Vehicles (Braitenberg), 324–25

ventromedial hypothalamus (VMH), 163, 182–83, 196

vinegar flies. *See* insects

vision or visual processing: "blindsight" and, 51; visual stimuli and, 59, 64, 162, 205, 207–8, **244**, 304–5. *See also* facial expressions

(VMHdm/c) dorso-medial and central portion of ventromedial hypothalamus, 183–86, 186–88

voxels, 237

Wager, Tor, 276

Wallace, David Foster, 153

What Emotions Really Are (Griffiths), 300–302

Winkielman, Piotr, 303, 306

worms *(C. elegans):* as model organisms, 114, 115, 126, 134, 198, 211, 214; roaming and dwelling states in, 211–12; simple nervous systems of, 198

Yarkoni, Tal, 231

Young, Michael, 199

zoom in / zoom out method, 39